MW00856323

"As an issue, noise often flies below our collective radar—in fact, it's often ignored completely. But *Clamor* grabs your attention from the very first word, telling a compelling and well-researched story about why we as a society ignore noise at our own peril."

—Rick Neitzel, professor of environmental health sciences, University of Michigan, and principal investigator of the Apple Hearing Study

"Sound powerfully affects our health for better and for worse. Pulling together the many influences noise exerts on life all in one place, *Clamor* will at last spawn much-needed awareness of this underrecognized pernicious influence."

—Nina Kraus, professor of neurobiology, Northwestern University, and author of *Of Sound Mind: How Our Brain Constructs a Meaningful Sonic World*

"Chris Berdik's *Clamor* comes at the perfect time. The 'appetite for sonic refuge,' as he puts it, has become a universal need in an era of ever-growing attacks on our basic senses. This book is a suitably calm and clear-eyed guide to the resistance."

—John Lingan, author of *A Song for Everyone: The Story of Creedence Clearwater Revival*

"*Clamor* is a wide-ranging survey of the diverse ways sound affects us, both bad and good. Our eyes may give us more detailed information about our surroundings, but what we hear affects us more quickly and—emotionally—more deeply than what we see, subliminally shaping our experience of the world. Chris Berdik's book will introduce you to the power of sound and the experts in many fields grappling with questions like what noise is, how much is too much, how sonic trash frays our attention, and how to design better alerts and even whole soundscapes to improve the quality of life. It will open your ears—and your mind."

—Dan Gauger, distinguished engineer
at Bose Corporation (retired)

"*Clamor* is not only a book about noise and how society reacts to it. Although it covers that ground, the most vital aspect of this book is where it leads: to a future where sound is imagined with intention and purpose, not just to a 'less noisy' din but to a better soundscape. Part history, part sociology, part scientific explainer, and at best a vision for our future sonic environment and how we will orchestrate it, *Clamor* is a must-read for anyone who cares about how their world sounds."

—Benjamin Markham, president of Acentech

ALSO BY CHRIS BERDIK

Mind Over Mind: The Surprising Power of Expectations

Clamor

How Noise
Took Over the World—
and How We Can
Take It Back

CHRIS BERDIK

W. W. NORTON & COMPANY

Independent Publishers Since 1923

Copyright © 2025 by Chris Berdik

All rights reserved
Printed in the United States of America
First Edition

For information about permission to reproduce selections from this book, write to
Permissions, W. W. Norton & Company, Inc., 500 Fifth Avenue, New York, NY 10110

For information about special discounts for bulk purchases, please contact
W. W. Norton Special Sales at specialsales@wwnorton.com or 800-233-4830

Manufacturing by Lakeside Book Company
Book design by Chris Welch
Production manager: Louise Mattarelliano

Library of Congress Cataloging-in-Publication Data is available.

ISBN 978-1-32400-699-2

W. W. Norton & Company, Inc., 500 Fifth Avenue, New York, NY 10110
www.wwnorton.com

W. W. Norton & Company Ltd., 15 Carlisle Street, London W1D 3BS

10 9 8 7 6 5 4 3 2 1

For Meaghan

CONTENTS

Part 2:
A Better-Sounding World

Clamor

PROLOGUE

Your chest tightens, your heart races. Your anger grows as your thoughts scatter. You can't focus on anything except for that sound—the incessant car alarm; the long, dull roar of a low-flying jet; your chattering office mates; the subway screeching into the station; or the rumble and rattle of a rooftop HVAC. But when the noise finally stops, you do your best to return to sleep, get back to your work, or resume your conversation. Your mind moves on to more important matters—at least until the next interruption.

This is the noise paradox. Sounds can trigger a visceral, even furious response from us in the moment but barely a shrug when that moment passes. When noise isn't in our ears, we usually don't consider it something to take seriously.

Overshadowed by other priorities, noise remains a problem regretted in hindsight rather than properly anticipated. Countless corporations have renounced walls and privacy partitions for a collaboration boost that often goes bust, while propelling noise to the top of the list of workplace complaints. Restaura-teurs eager for a modern look strip away every scrap of soft sound-absorbing material, plop kitchens into dining areas, and pump up the music until patrons must shout to be heard. Hospitals have amassed an array of patient monitors emitting alarm sounds that overload wards with frenzied beeps, freaking out patients and exhausting clinicians. Architects often treat acoustics as an afterthought, except in special cases such as concert halls, and urban designers typically consider sound only as far as regulations require.

The harms of this sonic neglect have been habitually undersold. Noise is pigeonholed as a nuisance or personal grievance, despite increasingly robust evidence that it's a serious and growing threat to public health. About 12 percent of American children and 17 percent of adults have permanent hearing damage from excessive noise—a loss that not only impoverishes our acoustic experiences but severs our connection to others. Hearing damage raises the risk of depression and dementia in older adults and can delay speech and language development in the very young.

Meanwhile, noise has also been shown to hinder classroom learning. In one famous study of a New York City elementary school located near an elevated subway line, reading levels of second- and third-graders on the building's train side lagged two to three months behind those of their peers in quieter classrooms; fifth- and sixth-graders on the noisy side were up to a year behind. Years later, after the New York Board of Education agreed to install sound-absorbing ceilings in classrooms and the transit authority added rubber pads to dampen the track noise, a follow-up study showed that reading scores had evened out across the school.

Noise can hurt us physically too. The daily (and nightly) din

is implicated in a long list of health ailments and diseases, ranging from hypertension to heart attacks. A 2020 report by the
European Environment Agency linked noise from transportation
and other sources to some 48,000 new cases of heart disease
and 12,000 premature deaths across the continent every year.
According to the American Public Health Association, noise could
be putting the health of more than 100 million Americans at risk.

This evidence, and the pervasive growth of the global din,
led the World Health Organization to count noise among the top
environmental threats to health. And as with most other pollutants, the mounting burden of noise falls heaviest on the most
vulnerable. That includes its clamorous advance into the natural
world, where it degrades both terrestrial and aquatic habitats.
The spread of noise and the dwindling of quiet have occurred
gradually, over many generations. In short, noise took over our
world while we weren't really listening.

How Did It Happen?

A big part of the reason for the ascendancy of noise is simple
math. Compared to a century ago, there are many more people
on the planet, living closer together and using a lot more sound-
emitting stuff. Globally, there are some 1.4 billion motor vehicles driving on roads that have carved up previously rural and
remote landscapes. The oceans are increasingly crowded with
massive vessels churning up an underwater cacophony as they
carry the fuel and freight of a global economy. Despite engineering advances that quiet the sound of jet engines, noise reductions
per plane have been more than offset by the burgeoning number
of flights passing overhead.

Two maps created by the Vermont-based nonprofit Noise Pollution Clearinghouse show the spread of noise across the United States over the past century—or at least the decimation of quiet. On the first map, a light-green tinge of "natural quiet" in 1900 fills most of the western states and covers vast splotches of the country east of the Mississippi River. Overlaying this light green is a smattering of dark-green dots, mostly in the Rocky Mountains and western deserts, representing the remaining vestiges of natural quiet at the outset of the twenty-first century. The second map shows the United States snarled in a thick web of black lines depicting the flights of commercial airlines, which blanket the nation in noise from above. Similar sonic dynamics have played out all over the world.

With the global population expected to hit nearly 10 billion increasingly urbanized souls by 2050, the math suggests an even noisier future. But not necessarily. There's much more to this predicament than demographics and the relentless march of technology. Our efforts to confront the acoustic fallout of these changes have been hamstrung by a myopic understanding of what noise really is, how it affects us, and what to do about it.

Noise is classically defined as unwanted sound, but over the past century, a single acoustic feature—the decibel—has become the canonical metric for judging what is noise and what is not. The decibel is a measure of sound intensity, a useful proxy for loudness.* Originally used in the late 1920s to quantify the loss of signal transmission power across miles of telephone cable,

---

* Loudness perception doesn't correspond directly to decibel levels. For instance, because human hearing is most sensitive to a middle range of frequencies, very high and very low frequencies must have more intensity to be perceived at a loudness equal to the middle range.

the decibel became the go-to measure of good and bad sound largely due to its simplicity. Decibels convert the immense range of sound intensity (measured by power over distance, or watts per meter squared) into something far easier to grasp. Consider that the faintest whisper we can hear has one-trillionth the energy of noise that can cause us pain—that is, the range from 0 to 120 decibels covers the twelve orders of magnitude between 0.000000000001 W/m2 and 1 W/m2. If you recall high school math, then you may recognize a logarithmic scale at work, such that a jump of 10 decibels equates to *ten times* more sound intensity. An additional 20 decibels is a hundred times the intensity, 30 more decibels is a thousand times the intensity, and so on.

While the decibel's utility is undeniable, it's only one measure of sound. Discerning noise by loudness alone is like the proverbial blind man trying to comprehend an elephant by the small bit of the animal he can grasp. The decibel's reign has given our judgments concerning noise a veneer of scientific objectivity, but it has also encouraged us to ignore vast realms of sonic perception. Reducing noise to loudness has undersold its true health risks and delayed a proper reckoning with its environmental toll. Just as important, if noise is defined only by loudness, then quiet must be the only sonic solution; thus less sound becomes the noise fighters' categorical imperative—a battle that, globally speaking, they cannot win.

Noise and New Sonic Frontiers

This book is a call to finally take noise seriously, not by complaining more stridently about it but by broadening our understanding of the problem and expanding the battle beyond decibel counts.

Only then can we devise smarter solutions to today's noise, better avoid noise in the future, and even strive to make our world sound *better*, not just less bad.

First, we'll explore the expanding universe of noise and its impacts, starting in the delicate spiral of the inner ear and radiating out to the human mind and body, and then beyond to our communities, and the planet. Loudness will never be irrelevant as we tour these realms of sound, but each step of our noisy journey will reveal sonic realities far more complex and nuanced than "too many decibels." The pages ahead will cover the tech, drugs, and rock and roll being marshaled to safeguard our hearing. They will tell tales of sound weaponized to liquefy our insides and the far less dramatic but likely far more potent dangers of sound-induced stress and broken sleep. We'll hear the fallout from the highways and heavy industry that are pushed through neighborhoods of least resistance to belch toxic noise, along with air and water pollution. We'll listen to human noise from nature's perspective and ask what nature has to say to us in return.

When writing about an underappreciated health hazard and inadequately studied pollutant such as noise, it's tempting to be alarmist. But, as we'll discover, there are lots of reasons for optimism. Expanding our definition of noise allows us not only to see new possibilities for acoustic fixes but also to understand what we might gain from taking all the sounds around us seriously, and not just the noise. In the largest sense, we will ask what it is that we really want from our soundscapes—the sonic kin to landscapes, encompassing the full mix of sounds in any given place. Do we simply want them to be less loud, or do we want something more?

There is tremendous value in quiet, and the appetite for sonic refuge grows in tandem with rising volume. In today's "silent airports," dueling gate announcements and garbled airport advisories have been replaced by improved electronic signage and notifications beamed directly to mobile phones. Some online real estate listings now display "quiet scores" alongside ratings for crime, schools, and walkability. Apps like SoundPrint, launched in 2018, offer crowdsourced guides to quieter restaurants, gyms, and other businesses, along with good spots for simply ducking out of the din. Nevertheless, while we may indeed want the world to shut up sometimes, we rarely want true silence. Just as problems with noise shouldn't be reductively defined as problems with loudness, neither should the solutions be limited to reducing decibels. Sound is not our enemy. As we'll see, it might even be part of the solution if our goal broadens from simply diminishing the decibels to cultivating soundscapes that work with us instead of against us.

With that in mind, we'll investigate the growing slice of our sonic world that is under our control, such as the sound of traffic on streets filling with electric vehicles. We'll examine efforts to transform the soundscapes of healing, learning, and working—including research into smarter alarms, which go beyond the beeps, and the art and science of *adding* sound to fight noise. We'll learn how architects and urban planners are partnering with soundscape experts and tapping into new technologies, so we can listen to a new office space, restaurant, or city plaza before it's built and make design choices with sound in mind. Finally, we'll explore the growing movement to forge new metrics and models that can compete with the decibel and even predict how people will react to soundscapes that don't yet exist.

This soundscape movement is still very young, but it's growing rapidly. In 2018, Berlin officials used an app to crowdsource citizens' suggestions for new "quiet areas" worthy of official protection, based on how they made people feel rather than how many decibels they racked up. That same year, Wales became the first nation to suggest that soundscapes be considered alongside decibel counts in future urban planning and development.

"Our lives are enhanced by conversation, laughter and cheering, music and the sounds of nature," a Welsh official wrote in the nation's noise-action plan, explaining the shift in focus. "A healthy acoustic environment is more than simply the absence of unwanted sound, and noise management must have a broader focus than simply clamping down on the decibels."

In contrast to the recent appreciation of soundscapes, anti-noise laws have existed for millennia, many recalling bygone struggles over sounds that have long since vanished from our world. Indeed, observed through the lens of history, fighting noise may seem like a quixotic quest. Even in the best light, the struggle is ultimately unwinnable, and at worst, it can appear to be less about the sounds themselves and more about who is making them, such as the noise complaints that often accompany a rise in immigration or gentrification.

We'll never be free of noise, and in a sense that's to be celebrated. Life hurtles along with all its intense subjectivity, challenging us with the messiness of dissent and difference. With that in mind, it's important to note that, even as this book digs into the real dangers of noise, it's not a guide to ridding our lives of all the unwanted sounds that we inevitably inflict on one another. Instead, we will dive deeper to explore how soundscapes shape our lives and how we, in turn, can shape them.

In the end, we can break through the noise paradox only by expanding our ambitions for sound. Doing so won't be easy, but I hope this book will convince you that it is well worth the effort.

PART 1

Noise from the Inside Out

1

HUH?!

HEARING CONNECTIONS

"Sound has power!"

The audiologist Deanna Meinke was holding a tuning fork in one hand, and in the other, a Ping-Pong ball dangling from a string—a representation of the inner ear. Her demonstration was well rehearsed: a professor of audiology at the University of Northern Colorado, Meinke is also the codirector of Dangerous Decibels, a nationwide initiative encouraging kids to safeguard their hearing.

Meinke struck the tuning fork against her desk and raised it slowly to the Ping-Pong ball, which sprang away, fell back, then jumped again, dancing spastically until the fork's hum faded. When she made her wide-eyed declaration that sound has power, her voice lingered on the word "power" as if casting a spell.

While sound is invisible and ephemeral, it is, fundamentally, a wave of energy triggering a cascade of molecular collisions. From the standpoint of physics, the drop of a pin and an earthquake are close cousins. Sound moving through the air at sea level cannot exceed 194 decibels, at which point the energy of the sound wave exceeds the atmospheric pressure, pushing the air rather than moving through it, producing a shock wave. In a denser medium, molecules smash together more rapidly, which is why sound moves through water about four times faster than it does through air. In the vacuum of space, by contrast, there are no molecules to vibrate, and silence reigns. As the famous tagline from the 1979 sci-fi classic *Alien* put it, "In space, no one can hear you scream."

When we think of noise primarily as loudness and define it by decibels, we are measuring the energy punch that the sound packs, regardless of its source, its meaning, or any other context. This punch can indeed be powerful, potentially damaging the inner ear's delicate anatomy and causing hearing loss. This alone makes exposure to noise a large and growing public-health problem. According to the Centers for Disease Control and Prevention (CDC), about 40 million American adults have noise-induced hearing loss, and by 2060, the number could top 73 million. Globally, 1 billion young people (ages twelve to thirty-five) are at risk of permanent hearing loss, and the World Health Organization projects that by 2050, about one in four people will have damaged hearing. Meanwhile, research increasingly suggests that noise starts to erode hearing well before standard screenings detect any deficit.

But the threat noise poses to our hearing is greater than degraded sensory perception. We hear with our brains as much as our ears, and the fallout of hearing trouble isn't trivial: kids who can't hear well in school are more likely to have delayed cognitive development and struggles with learning, while adults with hearing loss are at an increased risk of social isolation, depression, and dementia.

The deeper we venture into our hearing system, the more the problem of noise expands from decibels alone to something much more complex—something that harms the intricate connections between ear and brain in ways that are far more insidious than we realize. The damage can also profoundly affect our connections with one another.

Brute Force

Any exploration of noise must begin here—before a sound is noise, before it is even a sound. A pulse of energy passes through the air until it vibrates your eardrum, the translucent membrane at the boundary to your middle ear. There, three tiny bones known as the ossicles work like a kick-drum pedal to transfer that energy to a much smaller membrane at the entrance to your inner ear and the spiraling snail of fluid-filled bone known as the cochlea. Inside your cochlea, wavelets in the fluid triggered by the vibrations jostle thousands of "hair cells," which line the length of the spiral and are named for the tufts of hairlike stereocilia protruding from their tips.

Sounds range widely in frequency, or how quickly the sound wave oscillates—measured in peaks per second, or hertz. Human ears can detect frequencies between about 20 and 20,000 hertz; beyond that, we remain oblivious to a whole universe of sounds, such as Earth's deep geologic rumbles and the piercing screams of bug-hunting bats.

A hair cell bent by acoustic energy releases the chemical

neurotransmitter glutamate into neural connections at its base, which are known as synaptic ribbons. These synapses then fire off electrical impulses to the auditory nerve, which ferries the acoustic signals to the brain, allowing you to hear.

This is where loudness can start to cause trouble. Louder sounds carry more acoustic energy and trigger a larger release of glutamate. Too much high-decibel excitement causes hair cells to overfill their synaptic ribbons with glutamate, to the point where they swell up and pop—a grisly process known as excito-toxicity. Eventually, the battering caused by noise will kill off entire hair cells, but the first to go are these synaptic connections to the auditory nerve, disrupting the flow of neural messages between ear and brain, both upstream and downstream. For example, when you focus on a particular sound, the brain automatically dampens less-interesting frequencies by telling the corresponding hair cells to stiffen and resist the sonic waves pulsing through the cochlea. When inner-ear damage disrupts the incoming sound signal, neurons all along the auditory pathway try to turn up the volume by ramping up their firing rates. In relatively quiet surroundings, this flurry of compensatory neural activity keeps the perception of sound relatively intact—but the feedback loop can get out of hand, leading to the buzz of tinnitus or the sensitivity of hyperacusis, which makes ordinary sounds seem too loud and louder sounds painful to hear.

How much loudness is too much is a matter of ongoing debate. The National Institute for Occupational Safety and Health (NIOSH), for instance, suggests a limit of 85 decibels, which is as loud as heavy traffic, over an eight-hour workday. But that doesn't mean that 85 decibels is "safe"; it is simply the threshold that a fifty-year-old study had linked with a 15 percent rise in the risk of hearing loss for workers, compared to a 29 percent jump at 90 decibels.* Both the Environmental Protection Agency and the World Health Organization (WHO), meanwhile, recommend that people cap their exposure at 70 decibels (dishwasher loud) over twenty-four hours.

Digging a bit deeper, we find that the NIOSH guidelines cut the exposure time in half for every 3-decibel bump in loudness (a doubling of sound intensity). Thus, the recommended limit for 88 decibels drops to four hours and then to two hours for 91 decibels, and so forth. Likewise, the more conservative WHO suggests we endure less than one hour at 85 decibels.

Most of us have no idea how close we are to these decibel danger zones because almost all the loudness in our lives goes unmeasured. The best data comes from workplaces, and even that is quite limited. In the United States, NIOSH has traditionally focused on a handful of industries in which the most noise hits the most ears, namely, manufacturing, construction, and mining. There, workers' rates of hearing loss hover between 20 and 25 percent. In the mid-2010s, the agency launched a broad nationwide hunt for occupational hearing loss and found it everywhere, including among 19 percent of health-care workers and 17 percent of people in the service sector (a catchall category that includes workers in landscaping, law enforcement, and the food and beverage industry).

Even less is known about our exposure to non-occupational noise. Spot-checks of New York City's subway platforms, for example, found that trains screeching into stations emitted up to 111 decibels. In 2016, Boston researchers visited seventeen local

---

* In 1998, NIOSH confirmed the safety threshold of 85 decibels using a new model of hearing-loss risk that updated their excess risk estimates for 85 decibels and 90 decibels to 8 percent and 25 percent respectively.

spin classes, in which instructors blast music as they lead riders on stationary exercise bikes. The study found an *average* of 100 decibels there—a loudness NIOSH recommends we endure for no more than fifteen minutes without ear protection—and a peak of 116 decibels, which can damage hearing in seconds. A 2021 analysis of hour-long measures of sound levels at ten Nashville music venues found that they averaged 112 decibels and topped 101 decibels 90 percent of the time.

While the bulk of research on noise and hearing loss comes from Europe and North America, high decibels are a global problem, driven by the rapid rise of megacities in Asia and Africa, where industry, commerce, and hastily built homes are often jumbled together on overstuffed roads where the honking never stops. In fact, only one city in Europe or North America (New York) cracked the top-ten loudest in an analysis of traffic noise from more than sixty cities around the globe compiled for a 2022 United Nations report, which called these mounting decibel levels an "emerging issue of environmental concern."

Among the fallout from this growing din are rising rates of hearing loss. A 2020 study, for example, tracking nearly 72,000 Chinese workers in transportation, mining, and other industrial occupations, found that more than one in five had suspected noise-induced hearing loss. The same study included data from the Chinese Center for Disease Control and Prevention showing that confirmed cases of workers going deaf due to on-the-job noise leapt from 144 in 2001 to 1,555 in 2019. In India, meanwhile, physicians in the state of Kerala, having noticed an alarming increase in young people with hearing loss, founded the National Initiative for Safe Sound (NISS) in 2013 to hold conferences, raise awareness, and support local anti-noise activists. The doctors of NISS have also spearheaded studies that reveal an epidemic of hearing loss among traffic police and auto-rickshaw drivers in several Indian cities. From Mexico City to Accra to Hanoi, people have been flooding municipal offices with noise complaints. Many of these locales have had noise-control laws on the books for decades, but enforcement is a struggle. For example, it took years of litigation by private citizens in Mumbai to kickstart enforcement of India's national noise regulations passed in 2000, which established "silence zones" with strict limits on decibels (50 by day and 40 at night) around schools, hospitals, and places of worship. And still, the noise remains tough to tame. A 2017 study of "silence zones" in seven Indian cities found sound levels up to 77 decibels by day and 75 decibels at night, well beyond what regulations allowed.

Some loudness is unavoidable, some is forced upon us, and some we choose. But it all adds up. Your ears don't care if the decibels come from your job, your local bar, or your power tools. They don't care whether the music blasting from your earbuds is rock, country, or hip-hop. If the sound is loud enough, it will cause damage, and a tiny piece of your sonic world will be lost.

What Is Lost

While the numbers from NIOSH and other agencies that track hearing loss around the world are alarming enough, there's reason to believe that hearing loss is even more widespread, and its harms even more profound, than previously understood.

Audiograms have been the gold standard for diagnosing hearing loss for nearly a century. They work by pinging a person's ears with beeps of increasing loudness along a range of frequencies. The beeps start out soft and grow louder until the listener detects them. The adult threshold for "normal hearing" is 25 decibels, and hearing loss is defined as a "threshold shift" at any frequency where the tone must be turned up to 30 decibels or more to be heard. These shifts tend to start at the higher frequencies, around 4,000 hertz, which is the range in which consonants get their crispness.

"An audiogram is a very simple task," explains Charles Liberman, a senior scientist at the Eaton-Peabody Laboratory at Massachusetts Eye and Ear in Boston.* "Is there a stimulus there or not? It doesn't require discrimination and understanding." If we screened visual acuity the same way we do hearing, the eye doctor would ask if you saw anything at all on the next line down of the eye chart without requiring you to read out the specific letters or numbers. Most likely, fewer people would wear corrective lenses, but we'd be a menace on the highways.

As it is, most people don't complain to audiologists about missing particular pitches. They are instead distressed by struggles to converse with friends and loved ones in a crowded place. They miss out on the stories and can no longer engage in the banter. They're embarrassed by their need to constantly ask others to repeat themselves.

When people can no longer follow conversations, they start withdrawing from social activities and become increasingly isolated, a risk factor for dementia. A study out of Johns Hopkins School of Public Health, for instance, followed more than six hundred adults over twelve years and found that mild hearing loss doubled their chance of developing dementia; severe hearing loss quadrupled it. Meanwhile, a 2020 meta-analysis by Aus-

---

* Liberman was the lab's director from 1996 until 2022.

tralian researchers found that hearing loss also raised the risk of depression by nearly 50 percent.

Nevertheless, up to one in five people who report difficulty with hearing in a noisy environment have unblemished audiograms. In fact, groundbreaking research by Liberman and colleagues found that noise can cause profound inner-ear damage long before standard screenings find anything amiss, a discovery they dubbed "hidden hearing loss."

In their seminal study on the phenomenon, published in 2009, Liberman and Sharon Kujawa, Mass. Eye and Ear's director of audiology research, tracked the damage to the ears of laboratory mice caused by two hours of high-intensity noise (100 decibels). Over the next two years (the typical lifespan of a lab mouse), they repeatedly tested the animals' hearing with the mouse equivalent of an audiogram, using a tiny electrode to measure the brain stem's response to tones in 5-decibel increments. Crucially, they also used a new staining technique to keep tabs on the inner ear's hair cells, synapses, and nerve fibers tethered to the auditory nerve.

Superficially, the blast of noise seemed to cause only temporary injury. Twenty-four hours after exposure, the mice had experienced some hearing loss, but their brain-stem responses bounced back to normal after a couple weeks and remained stable. What's more, there was very little hair-cell death across a mouse's entire lifespan. In sum, the mice seemed to have taken the loudness hit and emerged unfazed.

But when Liberman and Kujawa looked a bit deeper and examined the synaptic ribbons and nerve fibers at the bases of the hair cells, they discovered that the noise had caused immediate and widespread carnage. Up to half of these critical links between the hair cells and the auditory nerve were gone. In other words, noise-blasted mice could still perceive sounds, but they had far less ability to transmit fine-grain sound signals to the brain. Mice that were spared the noise pummeling also lost cochlear synapses as they aged, but at a much slower rate than the noise-exposed animals did. For instance, eighty-week-old mice sustained about 25 percent synaptic loss, but that was only half the toll taken by exposure to just two hours of intense noise.

"There was striking neural damage, and it was permanent," Liberman summarized. "If fifty percent of your nerve fibers can be wiped out with a single noise exposure, that's bad."

Liberman has been checking human ears for the same patterns of loss. But making the jump from rodents to people isn't simple, and the research is slow. Human hearing is affected by myriad genetic and lifestyle differences; by contrast, the lives of lab mice are ultra-controlled. Besides, human inner ears can be closely examined only after death, when a person's remains are donated for scientific study. As of late 2021, he and other researchers had closely examined the ears of about two dozen people, mostly adults, but also a few children. They had no details about the deceased individuals' exposure to noise, except that none of them had ever been referred to an audiologist for hearing troubles.

During our interview, Liberman used a whiteboard to sketch out a few of their findings on a graph comparing hair-cell death to the loss of auditory-nerve fibers in human ears. As ears age, hair cell numbers gently decline, never exceeding a 15 percent loss until middle age, when the nerve connections plummet. By age sixty, no ear had less than a 30 percent decrease in nerve fibers, and the majority had lost double that amount.

Even that much damage can easily be missed by an audiogram, Liberman stressed. The ear would retain enough nerve connections to succeed at the super-simple low-decibel task of detecting a beep. But as the listening tasks grew more complicated, the hearing system would be overwhelmed. Unable to code the finer contours of incoming sounds, listeners would experience the sonic equivalent of a pixelated image. This loss of precision matters because our biggest real-world hearing challenge isn't perceiving sounds but filtering them. Just to chat with a friend in a busy coffee shop, our brains rely on minute acoustic clues to localize and separate all the sounds buzzing around us, so we can distinguish our friend's voice from two types of noise interference: energetic masking and informational masking. Put simply, energetic masking occurs when extraneous sounds drown out what we're trying to hear with brute force sonic energy, such as the milk steamer hissing away behind the counter, which makes our friend's question impossible to discern. Informational masking, by contrast, happens when other sounds clutter our attention, such as conversations overheard from elsewhere in the coffee shop.

As noise obstacles go, energetic masking is the more straightforward. Air conditioners, engines, and various hums, rattles, and rumbles sound nothing like a friend's voice. Until these extraneous sounds hit a certain intensity, our brains can ignore them fairly easily. Informational masking can be trickier because distraction caused by sound depends on a lot more than loudness (more on this soon), and misplaced attention compounds any loss in hearing acuity.

The problem of informational masking is especially thorny in group conversations, where our focus must rapidly switch from one speaker to another. Dubbed "the cocktail party problem," the challenge of multi-talker conversations has vexed hearing researchers since the British electrical engineer and cognitive scientist Colin Cherry first investigated it in 1953. In his experiments, Cherry asked people to recite one of two brief messages played simultaneously over headphones. Everyone struggled with this task, but most eventually managed it after several repetitions and with closed-eye focus. Listeners had a much easier time when the two messages were separated, with one playing in the left ear and the other in the right ear. When Cherry told people to focus on only one ear, they could rattle off that ear's message perfectly, but with one remarkable side effect—the same spotlight of auditory attention that made it easy to disentangle one ear's message effectively canceled out hearing in the other ear. Even though both messages were played at equal volume, listeners could not recall a single word of the sentences spoken into their unattended ears. It was almost as if they hadn't heard them at all.

These experiments show that our ability to perceive sounds is just one cog in the intricate machinery of hearing, albeit an essential one. Hearing's most important role—connecting us to other people—is also the most vulnerable to noise, causing harm that begins before anything surfaces on an audiogram and extends well beyond anything it can measure.

Restoration Quest

Noise-induced hearing damage is irreversible, for now. Scientists have been trying to regrow human hair cells since the late 1980s, when researchers discovered that "progenitor cells" (close cousins of stem cells) in chicken ears spontaneously transform into hair cells to replace those destroyed by noise. This finding triggered a wave of optimism. If chickens could do it, then why couldn't we? Follow-up studies showed that hearing systems in other birds, reptiles, amphibians, and fish pull off similar tricks of regeneration. Sadly, mammals proved to be the exception. Nobody knows precisely why mammalian ears lack this protection, but there may be a loophole. Subsequent studies have found that neonatal cells in the ears of newborn mice spontaneously regrow damaged hair cells but lose that ability after only a week of life, when these cells develop to serve fixed functions. Since 2011, about a dozen labs have collaborated on the Hearing Restoration Project, supported by the New York–based Hearing Health Foundation, to deconstruct the genetic and epigenetic ingredients that promote hair-cell regeneration in chicks, zebrafish, and baby mice and block the same processes in adult mammals. After more than a decade of work, their collaborative efforts have identified several key genetic pieces and many more candidates, but a pill that can repair noise-damaged ears remains many years away.

According to the consortium's scientific director, the Harvard neurobiologist Lisa Goodrich, "It's like having a big box of LEGOs. You can build a million things with all those LEGOs, and you can sort them by size, color, and shape. It's a complex problem, but I think it's a solvable problem."

Liberman agrees. In 2013, he cofounded a company called Decibel Therapeutics* to develop gene therapies to regrow hair cells, among other projects. Nevertheless, Liberman's role in the company was to focus on research into regenerating the synaptic connections, which he thinks will be solved first, because the rest of the inner ear's hearing machinery—the hair cells above

---

* Purchased by Regeneron Pharmaceuticals in 2023.

and the spiral ganglion neurons below—remains intact for years after these delicate nerve endings pop.

"It's like the microphone is OK, and the cable is OK, but just this little connector isn't working," Liberman explained. "I'm reasonably confident that someday soon, somebody is going to figure out how to get that fiber to send out a new connection, because it's not that far." How far? About 100 microns, or the width of a human hair.

In fact, Liberman and collaborators have already bridged the gap in a mouse's ear, thanks to signaling proteins called neurotrophins, which help neurons grow and develop their unique functions. In 2016, they regenerated synapses and restored hearing using a surgical procedure to smear neurotrophin-rich gel on the inner ears of noise-exposed mice. Unfortunately, the fix worked only if the neurotrophins were applied within a day of exposure to noise. "We tried waiting a week, and it doesn't work," said Liberman. "We have not yet cracked that nut, but lots of people are working on it."

As this research progresses, the market for devices that promise hearing help in noisy situations has exploded, thanks partly to a loosening of federal regulations. In 2017, after years of debate, Congress legalized over-the-counter hearing aids, breaking open an industry that had long been controlled by a handful of companies providing devices only by prescription. It would take several more years for the FDA to finalize the new rules on over-the-counter devices (in 2022), but the boundaries between hearing assistance and hearing augmentation were already eroding. Among established hearing-aid makers and audio and technology companies selling "personal sound amplification products" (PSAPs), also dubbed "hearables," the impending regulatory changes stoked a multibillion-dollar cauldron of marketing hype and fast-moving technology.

Many of these in-ear devices can be controlled by smartphone apps and toggled for phone calls, music streaming, or personalized hearing assistance, with a mix of amplification and "noise reduction" strategies to help a desired signal, such as your friend's voice, rise above the din. Their onboard algorithms can boost incoming sound waves that match the patterns of frequency and timing that characterize human speech, such as rapid changes in pitch and tempo interspersed with pauses. They also hush obvious background sounds, such as the drone of an air conditioner or any sounds louder than a preset decibel threshold.

Despite all the new features, the "brains" of these hearing devices are no match for our own. Even the best hearing aids still struggle mightily with the cocktail party problem. While rapid advances in artificial intelligence can help us distinguish one person's voice from another, hearing aids still can't get inside a listener's head to know, at any given moment, which voice is the signal and which is noise.

Clearly, the most effective antidote to noise-damaged hearing remains prevention. Like others who do research on hearing, Liberman doesn't know how many hair cells and synapses we could save if we took better care of our ears, but he always keeps some earplugs handy for using the snowblower, mowing the lawn, or working with power tools.

"They're so cheap and easy," he said. "It'd be crazy not to use them."

Learning to Love Our Ears

Yet judging by NIOSH numbers alone, protecting our ears remains a hard sell, despite decades of noise guidelines and awareness campaigns. The majority of America's noise-exposed workers don't wear hearing protection, and even in the top-three noisiest industries—mining, manufacturing, and construction—where employers are legally required to offer hearing protection, about a third of workers still say "No, thanks." Outside of work, the idea of hearing protection barely registers. A nationwide study in 2018 revealed that a mere 8 percent of Americans made an effort to safeguard their ears at loud concerts or sporting events.

"We protect what we value," said Deanna Meinke, the Colorado professor who directs the Dangerous Decibels program, "and I don't think people value their hearing consciously."

She and other audiologists suggest that this apathy about our ears is driven by two main factors. First, many people falsely assume that hearing loss is simply part of growing old, like graying hair and wrinkles. Why worry about the inevitable? In truth, hearing acuity *does* degrade naturally with age, but studies show that a lot of "age-related" loss is likely due to a persistent auditory assault over the years, rather than the years themselves. Second, hearing loss is usually incremental; hearing erodes gradually, as sharpness declines and tones get muddled. And when our auditory inputs falter, our brains do their best to plug the gaps with a mix of memory and prediction. This slow decay disguises the true extent of what we lose as hearing fades.

As a conversation starter, Meinke routinely asks people a deceptively simple question: what's your favorite sound? Ask somebody about their favorite color, and the answer will be automatic; we've had a ready response since we were toddlers. This seemingly trivial detail almost forms a part of who we are. But when Meinke asks patients, or audiences at her talks on noise, to volunteer their favorite sounds, people hesitate as if they've never considered the idea.

These inquiries started a couple decades ago, when Meinke was trying to cajole a patient who worked in a factory to wear the hearing protection his company provided. The worker was a middle-aged man, a soft-spoken veteran who figured that noise was just part of his job and not something to worry or complain about. He would frequently pull four or five extended shifts in a row, so he could then escape to the foothills of the Rockies for a few days of fly fishing.

Wearing hearing protection at the factory was voluntary, but annual screenings for hearing were mandatory. Every year, Meinke would tell this man that his hearing was deteriorating and he should do more to protect his ears, and every year, he would politely ignore her advice. Then, one time she asked him if he had a favorite sound. After a beat, he smiled and described the soft, high-pitched kiss of a trout taking the fly at the end of his line and the little splash that followed.

"I said, 'You will lose that sound. It won't be part of your life if you don't protect your ears,'" Meinke recalled. From then on, the man wore hearing protection without fail. "It was like finding the key to a lock," she noted.

Over the years, Meinke has amassed some 3,300 favorite sounds. A whopping 70 percent of kids pick animal sounds. Human sounds, such as singing or a child's voice, are also popular, as are nature's elemental tones, caused by ocean surf or trees rustling in the wind. Smaller percentages favor the sounds of musical instruments or the acoustic tokens of beloved pastimes, such as the purr of an outboard motor, the growl of a Harley, or the crack of a baseball bat. About 2 percent of people prefer silence, but they rarely mean the absence of sound.

"They say, 'Silence is when I'm skiing, and it's the end of the day, and everybody's already down the mountain, and I stop and listen.' Or 'When I'm home, my roommates are gone, the TV's off, and everything is still,'" Meinke explained. "It's more about their quiet time, somewhere removed from everyday life."

Meinke's experience illustrates how getting people to truly care about hearing *before* it fades will take a more powerful motivator than scary statistics and stricter regulations. It will require a shift in the social norms that swirl around hearing loss. Imagine a future in which hearing devices carry no more stigma than prescription glasses, in which awareness of hearing's value, and vulnerability, at every age is more widespread, in which nobody looks askance at concertgoers who don a pair of musician's earplugs before the show.

The line between risks we accept with a shrug and those we take pains to prevent is constantly in flux. More than one audiologist I spoke with compared hearing protection to sunblock. Both hearing loss and skin damage tend to sneak up on people; often, trouble is not noticed until middle age or later (the average age at which melanoma is diagnosed is sixty-five). Over the past half century, sunblock has gone from almost unheard of to omnipresent. Despite the fact that no law requires that it be used, many parents today won't let their kids step outside in summer without spraying or slathering them with SPF from head to toe, and sun protection has also become a premium feature in lip balm, skin care, and makeup.

Despite their patina of permanence, social norms can change rapidly. In the late 1980s, for instance, hardly any musicians or their audiences wore hearing protection—it simply wasn't rock and roll. That's when the punk-rock bassist Kathy Peck found herself needing hearing aids in her twenties and founded HEAR, or Hearing Education and Awareness for Rockers. Soon, some of the world's biggest rock stars started speaking out about the hearing loss and tinnitus they endured, thanks to their years onstage. Pete Townshend, guitarist for the Who and a tinnitus sufferer, kicked in $10,000 to HEAR's cause and appeared in the nonprofit's 1991 documentary *Can't Hear You Knockin'*, along with other stars: Debbie Harry, Ray Charles, Lars Ulrich of Metallica, and Mickey Hart of the Grateful Dead. Years of awareness raising followed, including a music-and-sound-based physics curriculum for high schools, public-service announcements made by celebrities, benefit concerts, and partnerships with audio industry giants, such as Shure, Westone, and Etymotic Research, to promote custom-molded hearing protection for musicians. Today in-ear devices that both protect musicians' hearing and provide them with a clear feed of their own audio are standard gear on stage.

Concertgoers need ear protection too—and, encouragingly, norms related to that issue are shifting as well. In 2002, HEAR successfully pushed to pass a city ordinance in San Francisco requiring larger clubs and music venues to make hearing protection available to patrons, and Minneapolis passed a similar ordinance in 2014. Elsewhere, offerings of ear protection have proliferated where no mandates exist. In 2016, Pearl Jam announced it would provide free hearing protection at every show on that year's tour, and the House of Blues started selling hearing protection at all locations and online. A 2015 study of about a thousand people attending concerts in Toronto found that those who went to venues that provided earplugs were more than seven times as likely to protect their hearing. Australian researchers have likewise found that simply offering hearing protection at a venue promoted its use as much as educating patrons about the risks of hearing loss. In 2020, a Toronto startup, WHUT?!, put earplug vending machines in about two dozen music clubs. In another sign of the times, a *GQ* headline from September 2021 announced, "The Hottest Thing to Wear to the Club Is a Pair of Earplugs."

Hearing education is another powerful normalizer. Music colleges offer a growing number of courses on hearing health, pushed by advocacy groups such as Healthy Conservatoires. Meanwhile, in K-12 classrooms, the Dangerous Decibels crew elicits the help of Jolene, a silver fashion mannequin decked out in a matching black-leather jacket and cap, a blue wig, aviator sunglasses, and a sound meter wired into her ear, to demonstrate the power of sound that headphones and earbuds can pump directly into our ears. Jolene has family all over the world, mannequins based on the nonprofit's "Jolene Cookbook" and dressed for raves, clubs, or music festivals.

There is also a growing focus on noise and hearing in the realm of personal health tracking, an expanding multibillion-dollar universe of wearable technologies that keep tabs on everything from sleep to stress to cardiovascular fitness. In the fall of 2019, Apple launched the Apple Watch series 5, which can monitor wearers' sound environments and ping them when things get too loud, suggesting that they limit their time in high-decibel contexts or wear hearing protection. (These nudges typically begin at 80 decibels, in line with WHO's "Make Listening Safe" recommendations, but they are adjustable.)

At the same time, the Apple Hearing Study began, led by principal investigator Rick Neitzel, a public health researcher at the University of Michigan and a leading expert on noise and health. The study aims to find out how loudly people play their personal audio devices, track long-term patterns of noise exposure, look for links between decibels and hearing and nonhearing health impacts (principally heart health), and learn what sorts of features and reminders on our devices work best to protect hearing. The study's subjects agreed to anonymously share decibel exposure numbers, hearing tests, and heart-rate data from their Apple Watches or the iPhone health app. The researchers planned to collect data through 2024 from about 150,000 people, randomly sorted into two groups—one would receive more notifications about high-decibel exposures, inviting them to review their data and take a quick hearing test after especially noisy weeks. In the spring of 2023, Apple released a few topline findings from data gathered up to that point, including the fact that one in three American adults, equating to roughly 77 million people, were exposed to more daily loudness than WHO recommended from environmental noise alone, not including any additional decibels coming through their headphones

While protecting our ears from loudness remains our best weapon against irreversible hearing loss, the motivation to do it will require a deeper appreciation of everything we are protecting. There is a link between defining noise as loudness and considering hearing loss only in terms of degraded sensory perception—this mechanistic framing undersells the threat. Hearing, like our other senses, connects us to our world and to ourselves, and severing these connections cuts deep.

Simply put, ear protection is pro-sound, something Meinke stresses. "Our intention is never to tell people to stop doing noisy things," she said. "We want to equip them with the strategies to make healthy choices." Sometimes that means using earplugs. Sometimes it's turning down the volume. Other times, it's knowing when to walk away. Meinke preaches flexibility. Some protection is better than no protection. Sooner is better than later, and it's never too late.

2

HEARING OURSELVES THINK

NOISE DISTRACTION

While there is so much noise in the world, the realm where loudness reigns supreme measures only about an inch and a half—the average length of an adult human cochlea. Within this tiny spiral, acoustic energy is all that matters. But the farther a sound signal gets from the inner ear, the less loudness determines its noise potential, and the more its impact shifts from hearing to thinking.

During every waking minute, our brains must sift through a barrage of sensory stimuli, separating what's important from what can be ignored. Much of this filtering is automatic, but it can become a struggle when we're trying to ponder or problem-solve.

Indeed, the anguish caused by distraction echoes deeply in the history of our concerns about noise. Popular etymologies trace the word "noise" to the Latin term for "nausea," implying a visceral distaste for, or even revulsion to, unwanted sounds. Digging deeper into linguistic history, we find that the ancient

Greek root for "nausea" links it to a specific source—seasickness. The word for "ship" is "naûs." The idea of seasickness conjures a mind unmoored, unbalanced, and disoriented—a feeling you may recognize from a typical day at the office, when you're staring at your computer, facing down a tight deadline, and trying desperately to think while phones ring and coworkers hop in and out of video conferences or chat about their weekend plans.

Concerns about indoor noise are often framed as battles over loudness ("Can you keep it down?!"). But sounds that can fracture our focus register all along the decibel scale. In fact, and counterintuitively, a problem with noise may get worse when the overall level of sound decreases. Sounds snag our attention far more easily in an otherwise quiet environment. When clear thinking is critical, a lone intelligible voice can hijack attention, while a livelier churn of background sound may actually bolster concentration. More fundamentally, equating noise with loudness underestimates the risks often prevalent in moderate-decibel domains, such as offices, schools, and hospitals. The distractions that can result are more than just annoying; they sap productivity, increase the number of costly mistakes, delay cognitive development, and pile on chronic stress.

Fractured Focus

Not surprisingly, the first studies of office noise and its detriments pinned the blame on loudness. In the late 1920s, an industrial psychologist at Colgate University named Donald Laird analyzed the performance of female typists working in his lab's sound chamber. Some were tested as they worked in quiet conditions, and others in noisy ones, while their breathing rates, stomach muscle contractions, and other physiological data were measured. Typists in the quieter environment worked faster, made fewer errors, and were less fatigued by their efforts. Laird explained the noise impacts as an innate fear response, akin to a chronic startle reflex. He later demonstrated this response for audiences by lighting off firecrackers next to a female assistant, who reacted to the explosions by squeezing an inflation bulb attached to a gauge he dubbed the "Tensionometer."

A more nuanced perspective finally emerged in the late 1950s and early 1960s when acousticians from Bolt Beranek and Newman in Cambridge, Massachusetts, paired sound measurements with extensive surveys of people in offices and other workplaces. They found that the biggest complaint related to noise was not about high levels of ambient sound but rather a lack "speech privacy"—on one hand, worrying about being overheard, and on the other, being distracted by the conversations of others.

The notion that crisply transmitted speech might sometimes constitute a noise problem was something of a revelation for many acousticians, who had spent their careers trying to improve the intelligibility of speech and scrub distortion and other interference. In light of their findings, the Cambridge acousticians proposed "speech interference level" as a new measure of indoor acoustics. This measure could be used to estimate preferred levels of background sound for a variety of places, ranging from libraries to large offices to maintenance rooms. Today acousticians employ a related measure, the "speech transmission index" (STI), which ranges from 0 (unintelligible) to 1 (perfectly intelligible).

What's more or less desirable on this scale depends on the listener's motives. When a discussion between nearby coworkers sounds garbled to you at your workstation, that low level of speech intelligibility is bad if you want to be part of the conversation, but it's good if you'd rather tune it out.

The stress caused when noise distracts us has less to do with an extended startle response, à la Donald Laird's firecracker routine; it's driven instead by the frustration of stolen focus. Plenty of sounds can distract us—from the wail of a car alarm to the plunk of a leaky faucet—but one of the most potent attentional lures is an intelligible human voice. For example, studies find that people trying to complete a task are *less* distracted by a conversation in a language they don't understand than by the same conversation spoken in their native tongue. At the same time, people are *more* distracted by overhearing one side of a phone conversation—the notorious halfalogue—because nothing hooks our pattern-seeking brains more than a puzzle to solve.

It's also worth noting that people perform worse on cognitive tests in the presence of halfalogues than they do when a full conversation is taking place, even though the latter contains more extraneous sounds that must be ignored. (It may go without saying, but people score best in a quiet room.) Research into halfalogues also shows that distraction is driven by "semantic mystery," not by the start-stop interruptions that characterize a half conversation. When voices are scrambled, whether in a dialogue or a halfalogue, enough to become unintelligible, the cognitive impact on those who overhear them is equivalent.

In sum, once we move beyond the cochlea and into the brain, noise becomes a much different, more complex problem—one that decibel counts often misconstrue. Loudness still matters, but it can no longer be considered the essence of noise. The realm of sound that causes unwelcome distraction is far murkier.

Noisy by Choice

A degree of distraction is likely inevitable anywhere that people gather to talk and think. But we have made the problem worse and multiplied our noisy predicaments by discounting the importance of our wider sonic surroundings. Look no further than the ascendancy of open-plan offices.

For all their shortcomings, the cubicle farms that arose in the latter half of the twentieth century offered some semblance of visual and acoustic privacy. But soon enough, the rows of drab, felt-covered partitions became synonymous with mindless, soulless corporate drudgery. By the turn of the twenty-first century, the rising tech giants, considered to be titans of innovation and the harbingers of a new economy, shunned cubicles in favor of lofty, sun-splashed workplaces replete with casual communal spaces meant to encourage collaboration. In these sparkling high-tech hubs, private space was almost anathema—something that would choke off the free flow of ideas and reinforce antiquated roles and routines at work. The few walls that remained were often made of transparent glass, despite putting people at risk of injury from walking straight into them. The potential noise that might result from this open-plan orthodoxy was met with a collective shrug, or even welcomed as a sign of creative buzz.

"The attitude was: We're inventing a new world, why do we need the old world," the architect Clive Wilkinson told *Fast Company* magazine in 2020. Wilkinson's firm oversaw the 2005 renovation of Google's headquarters in Mountain View, California, and his subsequent clients were eager for that Google look. "They were less sure about their own identity," he recalled, "but they were sure they wanted to be like Google."

By 2017, at least 70 percent of American offices were open plan in design, according to the International Facility Management Association, despite mounting evidence that the racket it produced drove employees crazy and hampered their job performance. Noise distractions quickly became a top concern among rank-and-file workers, but not so much for their bosses, who were more likely to occupy the private offices that remained. Only 39 percent of executives felt that workplace noise was a problem, according to a 2016 survey of about twelve hundred office workers spanning several industries, which was conducted by the global financial forecasting firm Oxford Economics.

"Unsurprisingly, then, very few companies have taken meaningful steps to address the problem," concluded a report based on the survey. "Noise is an afterthought in office construction, and executives overestimate employees' ability to drown it out with the tools available to them."

In fact, when Oxford Economics did a follow-up survey two years later, the portion of employees who said they were able to mitigate the distraction caused by office noise had fallen, from 41 to 29 percent. A third of those surveyed were donning headphones at work to block out extraneous sounds, and three-quarters of the respondents said they had to leave the office and take a walk to properly focus.

Since 1996, the Center for the Built Environment (CBE) at the University of California, Berkeley, has regularly surveyed people on the quality of their workplaces, asking about everything from cleanliness to the comfort of the furniture, and garnering more than 100,000 responses related to about 1,000 buildings. Bad acoustics is consistently the number-one gripe, and a 2005 analysis of the CBE's data determined that the dissatisfaction is more about a lack of speech privacy, not high levels of ambient noise. A belated nod to the massive problems created by noise distraction is the emerging market for a new office amenity—small soundproof rooms the size of overgrown phone booths, sold by startups with names like Zen Box, Thinktanks, and Hush Box, that workers use when they need to escape the din.

Despite employees' deepening discontent and the scramble for post hoc fixes for noise problems, the open-plan office trend continues unabated. The ideal of a workplace without walls, buzzing with the sharing of creative ideas, is powerful, even if the promised boost in collaboration never arrives. Studies of companies that switched to open-plan offices find that putting everybody in one big room is no guarantee of improved collaboration. It might even be counterproductive.

In 2018, for example, researchers at Harvard Business School kept tabs on about 150 employees from two "Fortune 500 multinational companies" that were transitioning from cubicles to open-plan offices. For several weeks before and after the changeover, employees wore sensors during workdays, which tracked their movement, location, and speech (the presence of it, not its content). In the end, changing to an open-plan office *decreased* face-to-face interactions among colleagues by about 70 percent, while increasing electronic communications, such as texting and emailing, by 50 percent.

More businesses are starting to recognize this mistake: both cubicle farms and open-plan offices force a single soundscape onto every worker and every type of work. One alternative is the so-called activity-based office, which downplays assigned workstations in favor of an office geography organized around particular tasks, such as collaborative workspaces, meeting areas, lounges, and quiet zones. An area reserved for brainstorming sessions can be more open and noisier, for instance, as long as sonically sheltered places elsewhere are available for focused work.

That said, one of the biggest reasons why noisy open-plan offices persist has less to do with aesthetic or collaborative aspirations and more to do with simple economics. Ultimately, knocking down walls and eschewing private space for workers allows companies to save money on real estate. Studies indicate that employers switching from private offices to an open plan can cram at least three times as many people into the same square footage. The rise of hybrid work arrangements in the wake of the Covid pandemic has done little to dull this space-maximizing imperative, nor the problem with office noise. Now a coworker engaged in a phone conversation or Zoom conference with a far-flung team can be even more bothersome, because the old hubbub caused by a fuller, busier office no longer masks the sound.

In the cold light of economic decision-making, it's easier to understand why workplace sound environments are routinely neglected. What do a bunch of employee gripes amount to on a corporate balance sheet? More broadly, what's risked by treating noise like the weather—a fact of life that everyone loves to grouse about before getting on with more important matters?

"Not much!" was the answer—until recently. But research on the costs of distraction has been gaining momentum. They include the dangers of texting while driving and addiction to social media, both of which shatter our focus. And there's mounting evidence that noise distraction is a serious hazard for people in a variety of different environments, including offices, schools, and hospitals.

The Fallout

A 2023 meta-analysis published in the journal *Management Quarterly Review* reached the unstartling conclusion that chronically distracted employees are bad for the bottom line. The authors examined forty-six empirical studies of office arrangements and determined that any real estate savings companies obtained from open-plan offices were swamped by a "productivity tax" as noise-harried workers grew increasingly frustrated, error-prone, and unmotivated.*

The business investments needed to attract and keep skilled, high-functioning employees far outstrip the cost of office space, the authors observed, and so it's shortsighted at best to jeopardize the former to save a few dollars on the latter. The analysis also provided ample evidence that noise distraction not only degrades workers' productivity and morale—it can also harm their health. After switching to open-plan offices, employees complained more about stress and fatigue. Likewise, previous research by Cornell University psychologists found that clerical workers exposed to three hours of recorded office noise, averaging only 55 decibels and peaking at 65 decibels, subsequently had elevated levels of stress hormones and were less motivated to complete challenging follow-up puzzle tasks compared to people working without the added noise.

Employees who switched to open-plan offices also took more

---

* They also included studies of activity-based offices, where the amount of distraction landed somewhere between that of traditional and fully open-plan offices, with much depending on how well the zones for different tasks were defined and managed.

breaks, and a parade of studies from Sweden, Germany, Denmark, and Norway have found that workers in open-plan offices take significantly more sick leave than those in more traditional office settings or those working from home. Admittedly, causation is hard to pin down in these examples because both germs and noise proliferate more easily in an open-plan office. Nevertheless, separate epidemiological studies involving thousands of workers have found links between self-reported workplace noise and more days spent sick at home, regardless of office type.

Offices are far from the only places where noise can set minds adrift. Studies find that when classroom noise goes up, test scores go down. While most of this research measures the noise problem in decibels, evidence suggests that loudness tells only half the story about classroom noise. A 2014 report from the National Academies, based on ten years of data from schools near forty-six airports, linked airplane overflight noise starting at only 55 decibels to lower reading and math scores. While lab studies have found that students' cognitive performance is hurt by droning noises from sources such as airplane and vehicle traffic, students exposed to the same decibel levels of recorded classroom noise, including intelligible speech, fare even worse.

Chronic interruption due to noise can be especially hazardous for developing minds, according to brain and behavioral research done at Northwestern University's Auditory Neuroscience Lab, popularly known as Brainvolts. The lab's leader, neurobiology professor Nina Kraus, and her colleagues have shown that noisy environments can undermine cognitive connections that lie at the heart of language learning and literacy, such as following the rhythms of speech and sounding out phonemes. In her 2022 book, *Of Sound Mind*, Kraus argues that noise not only delays reading proficiency but also hinders a wide range of learning because the brain's sound-processing networks are wired into broader cognitive abilities, such as attention and working memory.

"We need to be less cavalier about the day-to-day commotion that surrounds us in our raucous world," Kraus wrote, "for the sake of our *brains*."

A Canary in the Operating Room*

Though a degree of sonic chaos might be expected in a classroom full of restive kids, noise distraction can also sabotage highly trained professionals doing critical work, with lives on the line. Hospital operating rooms, for example, are not the hushed environments one might imagine them to be. In addition to the growing array of machinery, monitors, and other electronic tools crowding into these super-reverberant spaces, a surgeon's musical playlist accompanies most procedures. During both real and simulated surgeries, noisier operating rooms correlated with more communication errors, more (self-reported) distraction among clinicians, and more postsurgical complications, according to a 2021 review of research on the topic.†

At the vanguard of the fight against operating room noise are anesthesiologists like Joe Schlesinger at Vanderbilt Univer-

---

* This section heading borrows heavily from the title of MacDonald and Schlesinger's 2018 article, "Canary in an Operating Room," which is also cited in this chapter.

† Some hospitals have taken a cue from the aviation industry and experimented with "sterile cockpit" protocols. These discourage unnecessary conversation during a surgery's riskiest moments, akin to the hush demanded during an airplane's takeoff and landing procedures.

sity Medical Center. Acoustic vigilance is a critical part of their job, as they listen for any changes in the vital signs of an unconscious patient.

Schlesinger leads a lab of biomedical engineering and medical students, along with undergraduate neuroscience and music majors, working to mitigate noise problems in hospitals. One of their earliest investigations, from 2013, found that recordings of typical operating-room sounds made anesthesiologists slower and less accurate at the critical skill of detecting pitch changes in pulse-oximeter beeps, which indicate a drop in blood-oxygen levels and potential danger for patients.

"When that beep turns to boop, anesthesiologists have a kind of instant, emotive response," Schlesinger explained. Even minor delays in reacting to falling oxygen levels can cause serious trouble, the study noted, because most anesthesia-related injuries and deaths are not due to one catastrophic error, but rather to a series of small mistakes that "cascade into a serious event."

Once again, the harm caused by noise in the OR cannot be accounted for by decibels alone. Schlesinger's study found that clinicians' performance dipped even though the beeps of patient monitors were much *louder* than the distracting background sounds (80 versus an average of 67 decibels). Because of this, Schlesinger's initial follow-up study aimed to bolster clinicians' ability to block out distracting sounds. After a few rounds of multisensory training, during which anesthesiologists practiced separating simultaneous stimuli—an image and a tone—that competed for their attention, the clinicians were better able to detect the auditory alerts of falling blood-oxygen levels in the midst of noise. Schlesinger and his coauthors argued that this approach could be used in training programs for anesthesiologists across the country. "These findings," they wrote, "are a first step in a line of research that has the capacity to save lives."

But soon afterward, the research took a new turn when Alistair MacDonald, an anesthesiologist in Missoula, Montana, happened to read Schlesinger's second study. MacDonald had recently experienced a noise-related emergency that almost led to a patient's death. The near miss occurred when MacDonald was part of a team finishing an abdominal surgery. The incisions had been stitched up and dressings were being applied, but the operating room was anything but peaceful as it was rapidly being readied for the next patient.

Along with the usual hums and buzzes of fans, heaters, and surgical suction came the rumble of carts being wheeled in and out and the clang of metal instruments being tossed into trays. Most of the surgical team assumed the job was done, and so conversations proliferated as the surgeon's Spotify playlist blared.

MacDonald didn't join in the banter. For him, this was the hardest part of the procedure and required intense focus. He needed to gradually wake up the patient, so she could breathe on her own, while he removed the ventilator tube from her throat.

Trouble started as the woman emerged from the paralysis of sedation. Her cough reflex kicked in, and she fought against the breathing tube before her lungs were fully functioning. At the same time, MacDonald noticed the woman's skin was turning blue. The noise in the room had obscured the slight dip in the pitch of beeps from the vital signs monitor, but a glance at the display confirmed his suspicion: the woman's blood oxygen levels were plummeting. He had to act within seconds to save his crashing patient, but he couldn't focus. Finally, he shouted above the din, "Be quiet and turn off the music!"

In the sudden hush, MacDonald made a quick decision. Rather

than put the woman back under sedation, he removed the tube, gave her a small dose of anesthesia to ease her cough, and helped her along with a hand-powered bag ventilator until she could breathe independently.

In the aftermath of this incident, MacDonald dedicated himself to fighting noise distraction in the operating room, and he zeroed in on music. He didn't want to banish playlists from surgery. Both the medical literature and his years of professional experience had shown that music buoyed the mood and energy of a surgical team and that longer procedures endured in silence would be torturous.

But what if there was a way to automatically kill the music in an emergency when focus was critical? MacDonald started brainstorming ideas for a device that could interface with both the operating room's audio system and the patient's vital signs monitors. If there was trouble, the device would hush the music. The quiet would help everyone think clearly in the crisis, and also the sudden change in sound would itself serve as an alarm, like the needle scratching off a vinyl record, leaving the room in silence as a signal that something was amiss.

Over the next several months, MacDonald pitched his vision to medical device manufacturers, but nobody was interested. They didn't see any profit in messing with the music clinicians clearly loved. Yet MacDonald's discussions with fellow anesthesiologists at this time dredged up more stories of near misses related to noise; in one tragic case, a teenage boy died during elective knee surgery in Michigan. That surgery took place at the tail end of a Friday, and everybody in the operating room was chatting about the weekend and enjoying the music. When the boy's air tube slipped out, nobody noticed until it was too late.

MacDonald first heard about this young person's preventable death when his own children were teenagers, and the tragic story fired his earlier musings into a mission. If the big well-established medical-device makers weren't interested in his project, then he would have to do it himself. With the help of some open-source coding tools he found online, a team of computer science students at the University of Montana, and a dedicated local software engineer and tinkerer, MacDonald cobbled together a prototype of the device in 2016—a toaster-sized creation he called the CanaryBox, in honor of another quiet-based alarm, the proverbial canary in the coal mine.

After reading Schlesinger's study, MacDonald wrote a letter to the journal's editor, suggesting that the researchers were asking the wrong question. Instead of offering sensory-perception training to help doctors improve their focus in noisy operating rooms, shouldn't we first try to reduce the noise distraction itself?

In his response, Schlesinger essentially agreed. He reached out to MacDonald and traveled to Missoula, striking up a research partnership based on the CanaryBox prototype. In 2018, Schlesinger and MacDonald ran a pilot study. Twenty-nine out of the thirty surgeons at MacDonald's hospital who tried out the device for a day of procedures gave it high marks and said they'd use it again. Next up was a larger clinical study at Vanderbilt Medical Center to test whether the CanaryBox improved clinicians' reaction times to warnings from patients' monitors. During surgeries that took place with a CanaryBox that was either switched on or off, clinicians tapped a small keypad when they perceived a drop in oxygen saturation. Across seventy operations, anesthesiologists responded to alarms nearly 7 seconds faster when the CanaryBox was activated, according to results published in 2022.

"That's huge," Schlesinger told me, especially given the nonstop workload of surgical teams. "Over the course of a whole day working in a noisy, cognitively demanding environment, if you can ease some of that attentional load and reduce the cognitive burden, then that's a potential benefit to patient safety."

Meanwhile, MacDonald had been courting investors for the CanaryBox, working out the device's hardware and software kinks, and trying to secure enough early adopters in hospitals to build a business case that might woo an industry partner. In 2022, the German company Storz Medical agreed to partner with MacDonald and officially launched the CanaryBox within a suite of integrated operating-room technology.

More Sound, Less Noise

However, turning down the volume isn't always the best answer to noise distraction. Consider how much easier it is for a colleague's halfalogue to snag your attention in an office that is otherwise whisper quiet. By the same token, *adding* some extra sound to a room can help make stray voices less intelligible.

By the 1970s, a handful of manufacturers started selling sound-masking systems to offices. These systems are designed to enhance speech privacy by pumping out just enough sound energy to churn otherwise intelligible voices into a meaning-free background buzz, which is easier to ignore. The added sound is often mischaracterized as "white noise,"* but it is typically more targeted at the frequency spectrum of human speech, with sound levels tailing off rapidly above a few hundred hertz. Sound-

---

* Defined as noise with an equal loudness across the entire spectrum of audible frequencies.

masking systems must be carefully "tuned" to a space and occasionally adjusted to avoid making a noise problem worse.

Even though sound-masking systems have been around for decades,* the notion of reducing noise by adding sound is still tough to reconcile with the longstanding mindset that equates noise with loudness and considers quiet to be the only solution. By the early 2000s, for example, the burgeoning trend in open-space offices had triggered a wave of noise complaints among workers in Finland, a country with a special affinity for quiet. Finland's tourism board recommends the nation's remote forests and snowy tundra as hushed retreats for the noise-weary. Indeed, as their issues with workplace noise continued to grow, the quiet-loving Finns initially tried to solve the problem by stuffing more and more sound-absorbing material into offices and mandating that air-handling systems be made even quieter. These strategies often backfired when the resulting hush made speech-privacy problems worse.

Eventually, an acoustics researcher in the Finnish Institute of Occupational Health named Valtteri Hongisto tried a new approach. Hongisto began his investigations by giving people a series of cognitive tasks in simulated office spaces. His subjects completed the tasks either in silence or while recorded "distractor" sentences were played at a constant comfortable volume of around 50 decibels. Compared to outcomes from those who worked in silence, the presence of distracting voices made cognitive performance fall sharply. But the distractor sentences were played with varying levels of distortion, and performance started to recover when the sentences were sufficiently garbled

---

* Globally, sound masking is a $141 million business, expected to hit $200 million by 2030.

to score 0.5 on the speech transmission index. When intelligibility dropped to 0.2, a subject's cognitive performance was indistinguishable from that of someone working in silence.

After defeating distraction with straight-up signal distortion, Hongisto achieved similar results by mixing sound masking with more traditional anti-noise measures, such as workstation dividers and office furniture made with sound-absorbing materials. Hongisto bolstered his case with a handy new metric called "distraction distance," which considered an office's mix of barriers, sound absorption, and sound masking to estimate how far from one another coworkers needed to be positioned for speech intelligibility to fall below 0.5. At that level, workers could more easily ignore extraneous chatter. The shorter an office's distraction distance, the closer people could work together without losing focus.

Hongisto argued that offices should aim for distraction distances of less than 8 meters (26.2 feet). By 2012, he had helped get distraction distance accepted as a recognized measure of room acoustics by the International Organization for Standardization, a sprawling NGO headquartered in Geneva (the standard was updated in 2022). Then, in 2018, the Finnish Ministry of the Environment established the 0.5 speech-intelligibility rating as a required standard for assessing office acoustics and recommended sound masking among a mix of anti-noise measures. As of this writing, Finland remains the only country to legally mandate speech-privacy considerations in office design.

"Even though it's now an international standard, there are still acousticians who only want to measure sound attenuation and don't want to hear the words 'sound masking,'" Hongisto told me. "Putting more sound into a noisy room makes no sense to them, especially for people who always just want to deaden decibels as much as possible."

Whether they work in an office or an operating room, noise fighters everywhere face this kind of inertia, which can be hard to overcome, especially when decibels are moderate and the harms caused by distraction are more indirect than noise-induced hearing loss. Other priorities easily take precedence over noise when the offending sounds can't be pegged to a loudness threshold and the fix will be more complicated than the blanket solution of making things quieter.

Like Valtteri Hongisto, Alistair MacDonald was surprised by this inertia as he struggled to sell people on the CanaryBox, which he believed would offer a straightforward solution to a serious noise problem. In an ironic epiphany, he realized that noise distraction often proliferates because we're too distracted to deal with it.

"In trying to fix one small corner of this problem, I've learned some humbling lessons," MacDonald wrote in an unpublished essay he shared with me. "Our modern world is overwhelmed with noise, our collective bandwidth is limited, and we're all creatures of our ingrained habits, even when they are dangerous or inefficient."

3

FEEL THE NOISE

HOW SOUND GETS UNDER YOUR SKIN

In October 2021, the American Public Health Association revised its definition of noise from "unwanted sound" to "unwanted *and/or harmful* sound." The small but telling change was prompted by American policymakers' stubborn refusal to take noise seriously beyond the hearing risks posed by high decibels bombarding workers in certain industries. The statement announcing the new definition was titled "Noise as a Public Health Hazard."

Beyond brute-force loudness, the potential harms caused by "unwanted sound" often wash out in a tepid swirl of personal preferences and gripes. Semantically speaking, "unwanted sounds" might be annoying, inconvenient, and maybe even offensive to listeners with delicate sensibilities, but they present no more of a public health threat than the "unwanted emails" clogging our inboxes or "unwanted commercials" interrupting a favorite show. Expanding the definition of noise to include "harmful sounds" evokes a much bigger problem—one that exists independent of personal likes and dislikes and doesn't confine its damage to the inner ear. Harmful sounds not only shatter focus and hobble cognition, as we've seen; they can also penetrate the rest of the body and raise the risk for several serious and potentially fatal diseases and events, including heart attacks and strokes.

"Calls for action have gone largely unheeded," read the statement on noise. "This policy calls for national noise standards, enforcement, education, outreach, and action on noise as a public health hazard. They are long overdue."

Ironically, the potential for sound to wreak physical havoc on us is guaranteed to make headlines if the suspected cause is acoustic weaponry—though evidence for the existence of such weapons remains sketchy at best.* Meanwhile, noise poses far more widespread and better-substantiated threats to our bodily health, and these go largely ignored because the physical toll that noise takes is indirect, caused by chronic stress and broken sleep. As is the case with hearing loss and distraction, noise's power to harm our bodies is not determined by loudness alone. A sound's duration, timing, and meaning can matter as much as its decibel count, if not more, and our subjective responses to what we hear are intertwined with reflexive reactions deeply rooted in our evolutionary past.

---

* As they did in 2016, after American and Canadian officials in Havana, Cuba, suffered mysterious ailments blamed on "sonic attacks" by a shadowy enemy, until multiple investigations ruled out acoustic weapons as the culprit.

Toxic Tuning

The belief in sound's potential for destruction has a deep history, evinced by tales, ancient and modern, of sound used as a weapon: the Indian mantras of the Rig Veda calling on war drums to drive away enemies and misfortune, the shofar horns toppling the walls of Jericho, and the sonic blasters of science fiction. This enduring fascination does have a foundation in science—acoustic energy can directly affect our bodies through well-established pathways—but the intrigue has been magnified and sensationalized by dubious stories of rogue scientists and powerful militaries working to harness sound to cause death and destruction.

Among the most infamous of sound weaponeers was the French scientist Vladimir Gavreau, who built several massive whistles in the 1960s, which were not intended to be heard but to kill. Made with lengths of twisted metal pipe encased in concrete, the whistles emitted up to 160 decibels of infrasound, meaning sound at frequencies below what humans can hear. In his lab on the outskirts of Marseille, Gavreau and his colleagues wore white lab coats, thick rubber goggles, and sturdy earplugs as they blasted themselves with infrasound from 7 to 196 hertz. Gavreau later described the experiences as painful and "very nearly lethal."

"Everything inside us seemed to vibrate," he wrote. "Presumably, if the test had lasted longer than five minutes, internal hemorrhage would have occurred."

Gavreau's killer-whistle research was grounded in the phenomenon of resonance, whereby acoustic waves match the "natural frequency" of a system—determined by a combination of its mass, shape, and stiffness—and amplify the resulting vibrations, potentially to the breaking point. One well-known illustration of resonance is the wineglass shattered by a singer's loud and sustained high note. Human bodies are made of sturdier stuff, but their vital components also have resonant frequencies, mostly in the infrasound range. Lungs, for example, have a natural frequency of around 25 to 30 hertz. Eyeballs, meanwhile, start to shimmy at about 19 hertz, and 110 decibels at this frequency can make eyes twitch and conjure flits of color or shadowy shapes. Studies of intense infrasound, typically above 130 decibels, usually focus on chronic occupational exposures from industrial machinery, wind turbines, and propulsion systems at close quarters. This research has connected prolonged infrasound exposure with dizziness, headaches, nausea, and uncomfortable breathing, but nothing to suggest a potency anywhere near to what Gavreau claimed.

The authorities in Marseille eventually shut down the killer-whistle research after vibrations from the experiments reportedly shook the walls of nearby houses and led to incessant howling by neighborhood dogs. Gavreau passed away shortly thereafter, in 1967. Thanks to hyperbolic media coverage, however, the destructive potential of his whistles not only lived on but ballooned to doomsday proportions. The 1973 book *Supernature*, for instance, falsely claimed that the first person to test Gavreau's infrasonic blaster "fell down dead on the spot" and that an autopsy revealed that the man's internal organs "had been mashed into an amorphous jelly by the vibrations."

A couple of years later, no less an influential voice than the beat icon William S. Burroughs trumpeted Gavreau's research in a published interview with Led Zeppelin's guitarist Jimmy Page. Fortified by tea, cigarettes, and whiskey in Burroughs's New York City apartment, followed by enchiladas and margaritas at a

nearby taqueria, they discussed the Loch Ness Monster, communication with dolphins, and eventually, deadly sound. Burroughs
told Page that Gavreau's whistle was powerful enough to "kill
everyone in a five-mile radius." He then wondered aloud if just
a dash of this high-octane infrasound, blended into music, could
do more to blow the audience's collective mind than rock and
roll's usual dry-ice smoke and laser shows.

In the years that followed, a few bands claimed to lace their
music with infrasonic vibes powerful enough to trigger orgasms
and pants soiling.* But the only musician to test Burroughs's
brainstorm somewhat scientifically achieved far more hum-
drum results. In 2003, the English composer and engineer Sarah
Angliss led a team of scientists and musicians who squeezed a
subwoofer that could pump out infrasound between 18 and 19
hertz into a 7-meter sewer pipe. That spring, at two concerts in
London given by the Russian-born pianist Genia (who was in on
the experiment), the pipe was "played" during selected compositions by a team member who controlled the device remotely;
no one at the performance knew when it was activated. The
audience filled out questionnaires about their physical and emotional reactions to each of Genia's compositions, and reports of
"strange feelings," including goose bumps and rapid heartbeat,
indeed rose 22 percent when the infrasonic pipe played.

This was a far cry from liquefied livers, or even altered consciousness, but it did jibe with the feelings of unease and anxiety that both media reports and scientific literature often link
with exposure to infrasound. For instance, in 1999 residents of
Kokomo, Indiana, complained about a constant low rumble causing sleep disturbance, headaches, and nausea—one example of

---

* Notably, the British industrial music pioneers Throbbing Gristle.

the global aural phenomenon known as "the hum." The source of the Kokomo hum was a mystery,* but the mounting complaints and reports of illness raised a furor that prompted the National Institutes of Health (NIH) to investigate infrasound toxicology in 2011. After reviewing sixty-nine studies, including ten with human subjects, which covered everything from heart health to cognitive impacts, the NIH reached this rather unsatisfying conclusion: "Most studies reported some effects attributed to infrasound exposure, though many studies also reported no observable effects. Among the more consistent findings in humans were changes in blood pressure, respiratory rate, and balance."

Meanwhile, several of the world's most powerful militaries have doggedly pursued development of infrasonic weapons in the past half century, without much evidence of success. Infrasound is tricky to aim, due to its very long wavelengths (upwards of 50 feet). Plus, the power of sound dissipates as it travels through the air, so generating enough infrasonic decibels to make somebody uncomfortable, let alone incapacitated, from a reasonable distance would require massive speakers and enormous power resources. A 2007 assessment by US military forensics experts, published in *Military Medicine,* concluded, "Although high-intensity infrasound significantly disrupted animal behavior in

---

* Eventually, acousticians hired by the city zeroed in on two sources of intense infrasound—a cooling tower at the Daimler Chrysler plant and an air compressor at the Haynes International metal fabrication facility. After the companies took steps to quiet these sources, hum complaints decreased but didn't entirely disappear. Occurrences of "the hum" and related research are tracked globally by "The World Hum Database Project," https://www.thehum.info.

some experiments, the generation of such energy in a volume large enough to be of practical use is unlikely."

At the other end of the frequency spectrum, above the range of human hearing, is ultrasound, which at 140 decibels can start to heat skin and internal tissues. Intense ultrasound can also churn up microbubbles in tissues, and the shock waves created when they burst can cause cellular damage or, when properly targeted, break up kidney stones and blood clots. While easier to aim than infrasound, an ultrasonic beam is also much easier to evade because higher-frequency waves lose energy faster and are more readily absorbed by trees, walls, and other barriers.

That leaves us with audible sound, and sonic weapons using old-fashioned loudness aimed at the ears, notably flash-bang (stun) grenades and the sound cannon known as the long-range acoustic device (LRAD). Created for the military after the small-boat terrorist attack on the USS *Cole* in 2000, the LRAD can function as a souped-up bullhorn for issuing commands to disperse, or it can be used to inflict pain with a "deterrent tone." The military version looks like a gray-sheathed spotlight mounted on ships and Humvees that can hit targets with up to 162 decibels, but municipal police forces are increasingly using slightly lower-power LRADs to control crowds and disperse protestors.*

Nevertheless, speculation about secret sound-weapons programs continues, fueled not only by the scale of resources devoted to the attempts but also by the strong primal con-

---

* In 2021, the New York Police Department agreed to ban the use of the deterrent tone when settling a lawsuit brought by protesters and journalists, who claimed that being hit with a deterrent tone at very close range caused permanent hearing loss and tinnitus, along with migraines and vertigo. Both district and appeals court judges ruled that extreme loudness was a use of force, leading to the settlement.

nections between sound and our psyche. There is undeniable power in a loved one's voice or a sudden thunderclap. It's but a small mental leap to imagine that a derivative of that elemental potency could be devastating if harnessed and well aimed. And yet sounds need not be weaponized, nor especially loud, to affect us physically. While the mechanisms of harm are less direct than an acoustic blaster, noise is nevertheless implicated in a wide array of serious health ailments that do not stem from rattled skulls and vibrating viscera, but rather from the mind-body connection.

Serenity Now? Noise and Stress

Stress is a fact of life. It can even be helpful in small doses: it's vital to physical function and critical to surviving encounters with threats as large as a predator or as tiny as a virus. Chronic stress, however, is toxic. When the biochemical changes it triggers are never shut off, they can take a serious toll on every bodily system and potentially lead to a cascade of serious mental and physical maladies, from heart disease and stroke to depression, diabetes, and even cancer.

Evolution has equipped all of us with a hearing system wired first and foremost for survival, allowing us to detect potential threats from a great distance, even when they're concealed. Consequently, sound is a hair trigger for the fight-or-flight response of the sympathetic nervous system, registering in the brain much faster than visual stimuli do. Only milliseconds after hearing some new rumble or thud, for example, the amygdala atop your brain stem kicks off an automatic stress response. A split second later, your cortex gets involved—zeroing in on the sound's source, tapping into associated memories and meanings, and weighing your options—and either engages the calming parasympathetic response or ratchets up your stress.

Thus, noise stress is caused by a mix of automatic and learned reactions, and in both of them, context and subjectivity matter. While loud sounds can certainly be quite stressful, the degree of stress a sound causes is not determined by its decibel count. Consider, for example, the panic and rage that people with misophonia feel in response to everyday sounds like chewing and throat clearing and fidgety sounds like finger tapping or utensils clinking against a dish. This condition was first observed clinically in the late 1990s and named in 2001. "Misophonia" means "hatred of sound," yet people with this disorder don't dislike sound generally, nor do they possess super-sensitive hearing. Loudness isn't the problem; in fact, it can be their salvation, especially at mealtimes, when the din drowns out the sounds of eating.

People generally dislike the sounds that commonly trigger misophonia, such as sniffling and lip smacking, but most of us react to these with a shrug, considering them minor annoyances. By contrast, when people with misophonia hear these trigger sounds, their hearts race and their palms sweat, indicating a hyperactive fight-or-flight response. At the same time, this hypersensitivity is very specific. A 2017 brain-scan study by British neuroscientists found no differences in the brain activity of people with and without misophonia in response to neutral sounds, such as rain, or to unpleasant sounds, such as a scream of pain or a baby crying. Only the common trigger sounds made the brain of someone with misophonia react differently—generating a lot more action in the anterior insular cortex, an island of neural connections tucked beneath the frontal lobe, which is part of the brain's "salience network." This network sorts the stimuli that require attention from those that can be safely ignored.

It remains a mystery why any brain would attach an outsized salience to a slurp of soup but not to a sound with more obvious ties to danger, such as a person screaming in pain. However, most misophonia triggers share at least two particular attributes: they are usually made by people, and they tend to spark the worst reactions when listeners feel they can neither stop nor escape the sounds. To date, there are no evidence-based treatments for misophonia. There is instead a menu of coping strategies that work for some but not for others, such as redirecting attention by wiggling a pinky toe and relaxation techniques such as controlled breathing. By far, the most common coping strategy is avoidance. People with misophonia become practiced escape artists, not only leaving the dinner table but also fleeing jobs, shedding friendships, shunning social gatherings, and constantly rearranging their lives to avoid certain sounds.

While misophonia is an extreme example of a sonically stoked fight-or-flight reaction, we are all susceptible to chronic stress inflamed by the everyday din. The street construction lingering for weeks outside your windows, the overflights interrupting your conversations, or the growly buzz of a poorly designed rooftop HVAC robbing you of a moment's peace—all these exposures mix into the stew of stress stirred up by financial worries, family strife, work struggles, and other tribulations.

Increasingly, scientists are connecting the biological dots between noise exposure, chronic stress, and the physiological precursors of disease. Studies of mice and human volunteers have found that a chronic tickling of the sympathetic nervous system by moderate-decibel noise floods the bloodstream with the stress hormone cortisol and rampaging free radicals that damage the cells lining blood vessels, raise blood pressure, and constrict the cardiovascular system with inflammation.

In one such study, cardiologists at Massachusetts General Hospital (MGH) showed how stress from noise and other sources worms its way into the body and raises the risk of heart disease, the world's leading killer. They analyzed hundreds of full-body PET (positron emission tomography) scans of patients that were taken seven years apart. These patients had all been successfully treated for cancer, and the scans were part of routine follow-up screenings.

The doctors classified patients as "stress exposed" if they lived in a neighborhood with a lot of road or airplane noise or in an area designated high crime or low income. Over seven years, "stress-exposed" patients were nearly twice as likely to develop heart trouble between scans, including arterial inflammation and heart attacks, controlling for other risk factors such as age, cholesterol levels, body mass index, and family history of coronary disease. Interestingly, when it came to risk, the source of stress didn't matter—that is, the effects of living in a noisy place were similar to the effects of living in a low-income or high-crime neighborhood. The risk more than doubled for people exposed to multiple stressors. The scans, meanwhile, revealed that patients who developed heart trouble over the years were much more likely to have a highly active amygdala, as well as more activity in the spleen and vertebral bone marrow—two key sources of white blood cells that help the body fight off disease and can also contribute to artery-clogging inflammation when chronically triggered.

These findings dovetail with mounting epidemiological evidence linking chronic noise exposure with a higher likelihood of disease. For example, an analysis of fourteen studies from Europe, North America, and Asia found that the risk for heart disease rose an average of 8 percent for every 10-decibel uptick in highway noise. Another study linked a 10-decibel jump in nighttime airplane noise with 14 percent more hypertension among adults living near several major European airports, controlling for other risk factors such as age, alcohol consumption, and exercise habits. Finally, a 2021 study in *European Heart Journal* compared the relative contributions of several risk factors in 24,886 deaths from cardiovascular disease among people living near the airport in Zurich, Switzerland, and attributed 3 percent of the deaths to overflight noise.

Unsurprisingly, regulators such as the US Federal Aviation Administration (FAA) don't gauge noise harm by risk of disease but rather by the far more benign measure of annoyance. Specifically, they infer the degree of annoyance caused by airplane overflights by using a decibel-based metric known as DNL (day/night noise level), a twenty-four-hour average loudness, with a 10-decibel bump for night flights. For more than forty years, the FAA predicted noise annoyance with a model known as the Schultz Curve, developed in 1978 by the Boston acoustician Theodore Schultz, who used surveys to estimate how many people would be "highly annoyed" by airplane noise at different decibel levels. According to Schultz, most folks were OK with a twenty-four-hour average that stayed below 65 decibels, at which point only about 10 percent of people felt "highly annoyed." When aircraft noise exceeded that loudness, the ranks of the highly annoyed rose steadily, to about 35 percent of people at 75 decibels. Thus, a DNL of 65 decibels became the FAA's benchmark for overflight noise harm, as well as the contour line on the exposure maps created by airport noise models within which the agency agreed to pay for soundproofing on homes.

However, the 65-decibel threshold has long been controversial. For one thing, people don't experience airplane noise as a twenty-four-hour average, but rather as many events of different durations and intensities. In the decades since Schultz calculated his annoyance curve, better technology has made jets quieter, but there are a lot more of them flying overhead. Making matters worse from a noise perspective, the FAA introduced its GPS-guided NextGen navigation system in 2013, which reduced variation in takeoff and landing routes, thereby boosting air-traffic efficiency and lowering fuel use and emissions. But streamlining navigation also concentrated the roars and rumbles on those people living below the designated flight paths. Skyrocketing noise complaints have reflected a growing furor both within and well beyond the 65-decibel DNL contour line. At Boston's Logan Airport, for example, 2,331 complaints were submitted by 391 people in 2012, the year before NexGen was implemented. Within a few years, the number of complaints had multiplied by a hundredfold.

In response, the FAA sent out new noise-annoyance surveys in 2015 and 2016, eliciting more than 10,000 responses from people living near twenty different airports. The FAA used them to generate a new National Curve, unveiled in 2021. The new findings suggested that the Schultz Curve had seriously underestimated how much overflight noise bothered people. According to the latest surveys, about 20 percent of people were already seriously annoyed when the DNL reached 50 decibels, and by 75 decibels, nearly 90 percent reported annoyance. In the spring of 2023, the FAA announced that it would be reviewing its noise policy, including the 65-decibel DNL threshold, and the agency invited several months of public comment. Nearly 5,000 responses were received, but as of early 2024 no policy changes had been proposed, and the FAA did not respond to requests for comment.

Because no sound is inherently good or bad, it can be tempting to think of noise as a synonym for "sounds I don't like." Accordingly, it's commonly thought that a person who is stressed out by sounds is akin to somebody who dislikes the smell of potpourri, finds opera tedious, or recoils from the look of socks in sandals—understandable reactions, but ultimately a matter of personal preference. As we've seen, however, science suggests a more complicated story. Our reactions to specific sounds are subjective, but the experience of being stressed by noise is universal and underpinned by hardwired connections we all share.

When Night Fails: Noise and Sleep Disruption

The timing of noise massively influences the amount of harm it causes. Specifically, most of the noise-linked increases in heart attacks and disease risk spike when overflight or traffic noise continues at night, a time when stress and disrupted sleep feed on each other. Sleep studies, meanwhile, suggest that noise can splinter our slumber below the level of our conscious awareness, harming health even among people who claim not to be bothered.

Sometime tonight, perhaps after reading a few ponderous pages about sound and health, you will drift off to sleep. Your eyes will close, but your ears will remain open, listening for potential threats. The guard duty our ears perform automatically is particularly important at night, according to Mathias Basner, a psychiatrist and researcher of sleep and noise impacts at the University of Pennsylvania. From an evolutionary perspective, prolonged sleep is a necessary but dangerous behavior, because it makes us easy prey.

When Basner talks about noise disturbing our slumber, he doesn't simply mean the sudden startles that end with a reassuring pat from our partner or a shuffle to the bathroom. He means something much quicker, more frequent, and often below the radar of our awareness. Measures of brain and heart activity suggest that we have about twenty to twenty-five "awakenings" of at least 15 seconds' duration without fully regaining consciousness in an otherwise undisturbed night. This pattern is normal as we cycle between light sleep, REM (rapid eye movement), and deep sleep, and as our bodies shift positions.

"It's probably just a way for the body to check in briefly on the environment, making sure everything's still safe," Basner told me. Nevertheless, his research indicates that this constant vigilance has a downside when night noise ratchets up the number of these awakenings, resulting in serious consequences for health and well-being.

In 1999, after graduating from medical school in Germany, Basner parlayed a part-time job he took at a sleep lab to defray his tuition into a posting with the German Aerospace Center, which was conducting a major sleep study. Space may be silent, but spacecraft can be very noisy, filled with communicative machines and a bevy of clinks, clanks, and alarms, explained Basner, who has also studied sleep for NASA since moving to the United States in 2004. Poor sleep reduces cognitive sharpness and can lead to poor decision-making under stress, which makes it a serious safety concern during long space missions, such as those aboard the International Space Station.

A good night's rest can be just as important for the earthbound. Sleep is a critical time for your mind and body to reset—regulating your metabolism, bolstering your immune system, resting your brain, and soothing your body's automatic responses to stress. But unfortunately, noise is among the many reasons why our relationship with sleep is increasingly fraught. More than a third of American adults don't get enough rest, according to the CDC, which has deemed sleep debt a "public-health epidemic" for more than a decade. Meanwhile, surveys indicate that short sleep is a problem worldwide, and the global market for sleep aids is expected to grow from about $78 billion in 2022 to $131 billion by 2032. The array of sleep-centered products includes over-the-counter medications, weighted blankets, temperature-controlled mattresses, and dozens of apps for tracking slumber or inducing pre-sleep relaxation.

Basner chose "awakenings" as a just-right measure of sleep disturbance—more sensitive than the rarer instances of fully regained consciousness but more meaningful to our physiology than the hundred or so nightly "arousals" that last only a few seconds. Sound need not be very loud to jostle our sleep at this level. Basner's research has found significant increases in nightly awakenings from transportation noise starting at only 45 decibels.

In 2022, Basner co-authored a sleep study of seventy-two adults in which road, rail, and aircraft noise was piped in at varying decibels for eight out of eleven nights. In this study, the researchers monitored the brain activity of sleepers to track their "wake propensity," a measure of sleep depth and quality. The researchers found that intermittent transportation noise at 45 decibels led to much shallower sleep, during which wake propensity rose by nearly 30 percent.

Many people swear that night noise doesn't bother them, and some say that they're so accustomed to the city's sonic churn that they struggle to sleep in its absence. But Basner's research suggests that noise can harm your health even if you don't notice the disturbance. "People can wake up hundreds of times during the night without regaining consciousness," he said. "They'll say they slept great, but their sleep is severely fragmented."

The physiological fallout of these disruptions can be profound ramping up the risk for heart trouble and stroke. Your blood pressure and heart rate slow down when you sleep, but they spike during awakenings, and research suggests that blood pressure elevated by noise-induced sleep disturbances will stay high the next day.

"Think about the possible health impacts of that happening again and again over a period of years," Basner said.

Of course, correlation is not causation. To make a case for causal links between disease and chronic noise, such as overflights, researchers need large longitudinal studies tracking actual exposure to noise, sleep quality, and health outcomes over years. Most noise studies, however, rely on exposure models extrapolated from only a handful of actual sound measures, which often estimate a home's noise exposure based purely on its proximity to major highways and airports. Sleep studies, meanwhile, are largely confined to labs. The sensors required for monitoring brain activity, muscle movement, heart rate, and breathing are so obtrusive that researchers typically throw out data from the first night, when slumbering subjects are still getting used to all the dangling wires.

"It's terribly hard to prove these things," said Basner. Nevertheless, he and other researchers are getting closer. In 2019, his lab launched an ambitious effort to simultaneously capture real-world measures of nighttime noise and sleep quality. With funding from the FAA, they validated a simpler, much less invasive way to calculate "awakenings," using only two electrodes that subjects could put on themselves at home, to capture heart rate and body movement.

In 2020, the researchers started mailing out sensors, along with small noise monitors that could fit on a nightstand, to about four hundred people across the country, a small subset of thousands who responded to a survey asking about aircraft noise and sleep quality. After collecting five nights of data, subjects were instructed to mail the equipment back to the lab for analysis. Delayed by the Covid pandemic, the analysis requires a painstaking examination of second-by-second physiological data paired with noise recordings every night, a process that was still ongoing as of this writing.

Some anti-noise advocates have labeled noise as the next secondhand smoke, but that comparison gives Basner some pause. While both noise and secondhand smoke are noxious by-products of certain activities, and both can lead to serious health problems that we have been slow to recognize, tobacco smoke is directly toxic to health, while noise is not.

"It's very much different, in the sense that the risks associated with smoking and secondhand smoke are much higher than what we observe related to noise," Basner told me. Tobacco smoke causes cancer and emphysema that can kill you outright. Noise has more indirect effects on health, even if the end result can be just as lethal.

"At the same time, noise exposure is so ubiquitous and so much harder to evade than cigarette smoke," Basner noted. "Noise matters on a public health level because so many people are exposed."

We shouldn't need to oversell noise toxicity or indulge the overblown claims of acoustic weaponry to recognize noise as a serious threat. Chronic stress and sleep deprivation are massive and growing public-health concerns in their own right, and noise looms large in both issues. Indeed, both managing stress and getting enough sleep are recent additions to guidelines for protecting heart health, such as those promulgated by the American Heart Association.

If we are serious about improving sleep as a public health priority, then chronic noise should be a bigger part of conversations aimed at finding solutions, alongside the usual suspects, such as 24/7 connectivity, with our mobile devices glowing faintly by our beds. The same could be said about including noise in any serious reckoning with chronic stress. Doctors screening patients for heart-disease risk should consider noise exposure and other stressors, along with unhealthy diet, smoking, and a lack of exercise. In short, our assessments of health risk ought to include a lot more listening.

4

THE NOISE GAP

SOUND POLLUTION AND
ENVIRONMENTAL JUSTICE

If a chemical venting from smokestacks or leaching into our groundwater was linked to the same widespread health problems as noise is, would we dismiss it as an annoyance or would we take it more seriously? Whether noise is emitted from heavy industry, transportation networks, power plants, or other major infrastructure, we should consider it a pollutant. In fact, the US government once did.

In 1972, the fledgling Environmental Protection Agency under President Richard Nixon opened the Office of Noise Abatement and Control (ONAC). It operated for less than a decade. ONAC's funding was axed shortly after President Ronald Reagan took office, in 1981.* During its brief tenure, ONAC set noise standards

---

* Technically, ONAC still exists and retains its authorities, but with zero staff or funding. In June 2023, the nonprofit Quiet Communities and retired federal prosecutor Jeanne Kempthorne sued the EPA in federal court to revive ONAC's operations.

for things like air compressors, trucks, and motorcycles; funded noise studies; and disseminated educational materials on the health hazards related to noise, including stress and heart disease. In a post-ONAC America, noise is mostly handled locally. At the federal level a handful of agencies remain involved, such as the FAA and the FHWA (Federal Highway Administration), which pays for all those hulking highway noise walls.

The blow to anti-noise efforts from the demise of ONAC was not simply the loss of research and regulatory leadership but also the diminishment of noise in the public consciousness. Rather than a widespread pollutant potent enough to be targeted by the EPA, noise is considered a serious problem only for a handful of high-decibel industries. Otherwise, noise has largely been relegated to the realm of isolated nuisance disputes and personal annoyance gauged by noise complaints.

But noise pollution is not an individual problem. It's a societal burden. The harms to health caused by noise are a hidden cost of commerce and convenience, which drains billions from the nation's economy every year. And, as with most environmental pollutants, the least powerful and most vulnerable among us bear the brunt of this noise and pay the highest price.

This "noise gap" wasn't created by chance. Along with other environmental health disparities, it grew from the legacy of racism and the social inequities that concentrated polluting infrastructure in neighborhoods of least resistance. Despite growing recognition of sound pollution's toxicity, the same power disparities that helped create the noise gap complicate efforts to fight the growing din under the banner of environmental justice. Not only must noise compete with more pressing hazards, such as unsafe drinking water and dirty air, but crusades for quiet can be hard to mount in communities that have historically been silenced, threatened by gentrification, or plagued by blight.

Sonic Injustice

About once a minute, the global economy roars past Thomasenia Weston's modest brick house on Livernois Street in Southwest Detroit. All day, and into the night, a cavalcade of eighteen-wheelers, hauling tankers and shipping containers for international logistics firms such as Hapag-Lloyd and HMM, roar and rumble past her residence. The trucks emit billowing clouds of soot, darkening windowsills that once were white.

"I don't normally sit out here. It stinks," Weston said, as we settled into folding chairs on her front porch one warm summer morning. She glanced at her dirty windows. "I just power-washed those last month."

Southwest Detroit, often known simply as "Southwest," is among the city's lowest-income neighborhoods. It's home to about 45,000 people, the majority of whom are Hispanic; about a quarter are African American, like Weston, who has lived here for twenty years, raised two kids, and serves as guardian for two grandchildren.

The predicament Weston and her neighbors face is far from unique. Across the United States, transportation and industry noise disproportionately burdens those who are least able to escape the din and have less green space to buffer it. In one of the first rigorous reckonings with our national noise gap, a 2017 study by public health researchers from the University of California, Berkeley, and Harvard found that the richer or whiter a census tract grew, the quieter it became. A 2020 audit of transportation noise nationwide uncovered similar inequities, as have local noise-exposure studies from cities such as Minneapolis, Atlanta, and multiple cities in California. Meanwhile, an analysis of environmental noise occurring near some 94,000 schools located throughout the United States found that those with higher proportions of students who were nonwhite or whose families had lower incomes endured far more noise from traffic and overflights.

Back in Southwest Detroit, Livernois Street is two lanes divided by a dashed white line, and it's clearly not built for heavy truck traffic. The street is chockablock with houses much like Weston's, with its chain-link fence enclosing a small front yard where a lone rosebush defies the crabgrass. Yet the neighborhood is sandwiched between interstates, freight distribution centers, and railways. It's also only a couple of miles from the Ambassador Bridge over the Detroit River, a major trade artery between the United States and Canada. As a result, Livernois and nearby streets are favored cut-throughs for trucks looking to bypass the clogged highways.

Commercial truck traffic between Detroit and Canada has grown steadily since the North American Free Trade Agreement took effect in 1993, rising to about 10,000 daily border crossings, which spill onto the residential streets of Southwest. A recent neighborhood count of trucks, taken by local activists and public health researchers at the University of Michigan, tallied about one commercial vehicle per minute at multiple locations. The same research partnership spawned a 2021 study of truck noise in Southwest, which tracked data from more than twenty locations around the neighborhood over several days. They found that daytime noise at sites facing the street sometimes topped

100 decibels and averaged around 75 decibels, about the same loudness as a vacuum cleaner. The noisiest sites were nearly 20 decibels louder than the three quietest ones, which were tucked in backyards away from the busiest roads; they were also only about 5 decibels quieter at night.

This pattern of noise concentrated with other types of pollution is no accident. It can be traced partly to discriminatory lending practices established by a federal agency known as the Home Owners' Loan Corporation (HOLC) in the 1930s. Using color-coded maps of about two hundred American cities, HOLC graded the risk of investing in different neighborhoods based largely on their racial and ethnic makeup. Areas where upper-middle-class white people lived were graded A, "best," and shaded green. B denoted "still desirable" areas that were colored blue, and the yellow shading of grade C characterized an area as "definitely declining"—here, residents were more likely to be working class or recent immigrants. Finally, places like Southwest Detroit, with growing numbers of people of color, were "redlined" and graded D, "hazardous."

Redlining led to disinvestment, and this in turn concentrated poverty, which sapped local resistance to new highways and other polluting infrastructure that harmed health. Redlining invited not only pollution but also bulldozers, which wiped out certain communities in the name of urban renewal. In the 1950s, for example, Detroit authorities demolished two neighborhoods at the heart of the city's African American life—Black Bottom and Paradise Valley—to make way for Interstate 375 and a public housing development.

Redlining was banned in 1968, but the vicious cycle it triggered continue to steer noise and other pollution into the most vulnerable communities. Multiple studies have shown that neighborhoods saddled with a lower HOLC grade in the 1930s have more pollution and less green space today than other neighborhoods with similar levels of employment, median income, and education. In Detroit, a 2022 analysis by University of Michigan researchers found that residents of the city's formerly redlined areas are about twice as likely to live within 5 kilometers of an industrial site registered as potentially toxic with the EPA, compared to non-redlined neighborhoods, while enduring 66 percent more "hazardous" road noise ("hazardous" meaning above a daily average of 70 decibels).

Of course, noise is rarely a stand-alone pollution. The aforementioned Detroit study, for instance, also found that people in formerly redlined areas breath air polluted with 12 percent more particulate matter from diesel exhaust—tiny bits of soot small enough to lodge deep in the lungs, which are linked with cancer, heart disease, asthma, and other respiratory conditions. A staggering 14.6 percent of children in Detroit have been diagnosed with asthma, compared to 8.4 percent statewide.

Weston has asthma, as do her grandchildren, who require nebulizer treatments. She keeps the street-facing windows permanently shut, which offers some protection from the dirty air. Still, the single-pane windows can't keep out the noise—the growling engines, the booms of empty containers hitting potholes, and the stuttering wheeze of hydraulic brakes as trucks near the stoplight at the end of her block.

"I don't sleep," Weston told me. Thanks to the constant racket, her grandchildren can't focus on their homework, and neighbors blame the heavy vibrations for cracking their front stoops and shaking loose mortar from their foundations.

Several times during our conversation, noise forced Weston

to stop talking and wait for a truck to pass by her house. At one point, after three trucks roared by in quick succession, she grew impatient and started shouting. Then she laughed.

"I have to laugh about it," Weston said. But she had also started fighting. A few years earlier, she had joined a nonprofit called the Southwest Detroit Community Benefits Coalition, which formed in 2008 when plans for the new Gordie Howe International Bridge were announced. It was slated to land right across Interstate 75 from where we sat.

The new bridge was scheduled for completion by 2025. Proponents promised it would bring thousands of new jobs to both Detroit and Windsor, Ontario, across the river, along with billions of dollars in regional economic gains, thanks to reduced congestion and more efficient trade. Of course, there would also be costs—in displacement and pollution, including noise from the crush of cross-border traffic that was, by some estimates, expected to more than double within three decades. All these burdens would fall heaviest on the people of Southwest and Delray.

To Weston and her family, the truck fumes were poison, and the incessant racket was a slap in the face—a loud, unrelenting reminder that their neighborhood had to put up with things that would never fly in a wealthier, whiter part of town.

"I'm going to go ahead and say it," she told me. "This is a Black and Brown neighborhood, and they feel like they can do whatever they want, and we'll just take it, take it, take it."

Sound and Fury

As more communities confront the noise gap, sound pollution is an emerging battleground for environmental justice. One front line in the fight against noise is Seattle's Beacon Hill neighborhood, a working-class enclave of about 30,000 people, more than 75 percent of them nonwhite, including many immigrants from Asia and Latin America. This long, skinny slice of the city is about 6 miles from north to south and a mile or two wide. Bordered by Interstate 90 to the north, Interstate 5 and King County International Airport (Boeing Field) to the west, a four-lane boulevard and light rail system to the east, and Seattle-Tacoma International Airport (SeaTac) to the south, it's hemmed in on all sides by noise.

Beacon Hill is directly below the busiest flight corridor for SeaTac arrivals, which means a daily parade of jets passing a few thousand feet overhead. By 2018, hundreds of flights a day were roaring over the neighborhood, and SeaTac was moving forward with plans to massively expand operations over the coming decade. That's when residents decided to organize, led by two local groups—the social service nonprofit El Centro de la Raza and the volunteer-run Beacon Hill Council.

They drafted a community action plan calling for changes in flight patterns, soundproofing mitigation, and tree planting, among other ideas, including their own program of on-the-ground neighborhood noise measurements, which kicked off that spring. They wanted a data-backed counterargument to SeaTac's noise model, which placed Beacon Hill just beyond the 65-decibel DNL (day/night noise level) exposure boundaries but used data from only one permanent sound meter in the neighborhood and a second just to the east.

Studies had already suggested a range of harms to people from overflight noise below 65 decibel DNL. For instance, five Beacon Hill schools were in the aforementioned National Academies study that found lower reading and math scores in schools situated near airports, starting at exposures as low as 55 decibel DNL. That was the same noise level at which heart trouble started to rise in a study of 6 million people over age sixty-five living near major airports around the world, including SeaTac.

But Beacon Hill residents wanted to go beyond abstract models of noise to measure real-world sound and to more accurately capture what it's like to have one jet after another fly over your house or classroom.

"People don't experience noise as an average, they experience it in spikes," said El Centro de la Raza's environmental justice coordinator, Maria Batayola, "and whether you're used to it or not, that noise triggers stress." So, with EPA funding and guidance from public health researchers at the University of Washington, a team of Beacon Hill volunteers and undergraduate interns fanned out across the neighborhood, taking measurements of sound at fifty-two sites over multiple days in the spring and summer of 2018. They found loudness everywhere. But unlike the neat noise-exposure boundaries suggested by SeaTac's noise model, the numbers on the ground varied widely. Sites registered between 58 and 80 decibels DNL, and the median was 68 decibels. During daylight hours, an airplane was clocked flying over the neighborhood every one to three minutes.*

As important as exposure data is, it tells only half the story of noise pollution. The other half is about the vulnerability of the people exposed. For instance, overflight noise in an affluent neighborhood is more likely to be buffered by ample tree canopy, higher-quality housing, and amenities such as central air conditioning, so windows can stay shut in warmer weather. The jets

---

* Unlike the models of airport noise, the Beacon Hill study did not single out airplane noise for the DNL calculations, but it did show that airplanes were the dominant source of noise at most sites.

flying over Beacon Hill, by contrast, add to the din from the surrounding highways and a host of other non-acoustic stressors—
both chemical and societal. The neighborhood is part of Greater
Duwamish, an area of the city that hugs a 5-mile stretch of the
Duwamish River, a polluted industrial estuary and an EPA
Superfund site. Data from the Washington State Department of
Health reveals that the Duwamish neighborhoods have higher
rates of poverty and lower life expectancy compared to more
affluent areas of Seattle and the suburbs.

As a result, Batayola and her allies haven't restricted their
fight to decibels. They're engaged in a broader environmental-
justice campaign to make Beacon Hill a healthier place to call
home. Besides noise measurements, they're gathering data on
particulates and carbon emissions from both aircraft and highway exhaust. They're also pushing for more trees to be planted
in Beacon Hill, one of Seattle's least green neighborhoods. In
2020, they partnered with the city's parks department and
raised money to purchase and preserve about twenty wooded
acres buffering the neighborhood from Interstate 5.

SeaTac, for its part, has recently stepped up its anti-noise
efforts, several of which were initiated by recommendations from
a bimonthly roundtable dialogue with local residents and businesses that began in 2018. For instance, in 2019, the airport limited nighttime use of the runway closest to local residences and
began to publish data on night-flight noise emitted by every air-
line. The following year, they accelerated the installation of sound
insulation, including window and door upgrades, expanding the
effort to apartment buildings, schools, and houses of worship, and
they began using a new "equity index" mapping tool launched in
2021 by the Port of Seattle, which includes SeaTac, to prioritize the
most vulnerable for new noise-mitigation projects and upgrades

to older insulation that no longer keeps out the noise, albeit only within the 65-decibel limits of its noise model.

Joining with other community groups under the slogan "Fix the Harm," El Centro de la Raza has pressured congressional representatives (thus far unsuccessfully) to push for expanding the funding of overflight noise mitigation beyond the 65-decibel perimeter for neighborhoods highly impacted by multiple types of pollution. The noise, the air pollution, the loss of trees, and the long-suffering river cannot be treated as separate causes, according to Batayola. "We want our health," she said. "It's all intertwined."

The Trouble with Quiet

Despite the urgency of combating noise pollution, doing so is far from simple—especially in communities that have had to fight to be heard. In the broadest sense, battles over noise are about power and who controls our sonic surroundings, which can make noise tricky terrain for environmental justice. A rise in noise complaints has often been a calling card of neighborhood gentrification and a harbinger of displacement. Studies have found that calls to the police non-emergency line (311) concerning noise tend to rise as neighborhoods become wealthier and whiter. These analyses can't discern the race of the callers nor the people making the noise in question. Nevertheless, research in 2016 by sociologists at New York University and the University of Cologne in Germany analyzed nearly 8 million geo-coded 311 calls and found significantly higher rates of complaints about noise and public drinking in census blocks within the "fuzzy boundary" between more racially homogenous areas.

In 2019, residents of a new luxury apartment building in a rapidly gentrifying section of Washington, DC, complained about a nearby store whose outdoor speakers played go-go music, a subgenre of funk layered with Latin American percussion. Go-go was born in DC in the 1970s—the decade the District earned the nickname "Chocolate City" due to its majority African American population. In go-go's heyday, live performances by bands like Chuck Brown and the Soul Searchers, Trouble Funk, and Rare Essence were the genre's beating heart. Go-go has struggled in recent years, however, because both DC and the music business have changed. With the proliferation of new development catering to more affluent residents, night clubs and other performance spaces that had featured live go-go were closed or displaced, including several venues located on the same street as the store targeted by noise complaints.

When police arrived at the store to measure the decibels, they found them to be within legal limits, but the store's owner nevertheless agreed to bring the speakers inside, scrubbing one of the last vestiges of go-go from the streetscape. That's when the protests started. People marched for the return of the music, and an online petition titled "Don't Mute DC's Go-Go Music and Culture" quicky garnered tens of thousands of signatures. After successfully bringing go-go back to the storefront, "Don't Mute DC" morphed into a movement to support go-go musicians, revive the genre,* *and* fight funding cuts to schools and social services in the District.

---

* Among other accomplishments, the group successfully lobbied Mayor Muriel Bowser to declare go-go the official music of DC in February 2020 and fund the music's preservation and promotion. They also helped raise money for a Go-Go Museum and performance space project that broke ground in the spring of 2024.

Worries about displacement and control of the sonic environment inevitably weigh on anti-noise efforts in gentrifying neighborhoods, including places like Washington Heights and Inwood on the northern tip of Manhattan. The leader of the community's noise task force in Washington Heights is Tanya Bonner, a middle-aged Black woman who's lived in the neighborhood since 2006. She has been called a gentrifier more than a few times.

"I get it. I can understand where it's coming from," Bonner said of the accusations. "It's defensive. People feel under threat."

Much like Washington Heights, Seattle's Beacon Hill is a neighborhood under pressure, beset by rising rents that have pushed out some longtime residents and businesses. As Batayola of Beacon Hill's El Centro de la Raza put it, "One thing I've really learned from this noise work is that when you're dealing with a community that you love, you don't want to diss your home."

Not that El Centro de la Raza should need to burnish its social justice bona fides. The nonprofit traces its roots to a 1972 sit-in protesting the defunding of ESL (English as a second language) classes and adult education offerings in local schools. In the decades since, the nonprofit has provided early education, food assistance, and workforce development, and it has hosted community meetings and cultural events. In 2016, the group opened a 112-unit building offering affordable housing, and they planned to redevelop another site with 87 more affordable apartments. All that hasn't stopped its anti-noise campaign from sparking some "love it or leave it" pushback.

"Every so often, people will try to shut us down by saying, why don't you move?" Batayola said, "And I think, wow, a lot of these families were here long before all the airplanes came."

Recognizing these complexities, Erica Walker, an epidemiologist at Brown University and leader of the Community Noise Lab, has developed an approach to pollution research that puts the communities themselves in charge. Walker has long understood that decibel counts and exposure models can't capture the full reality of noise and how people experience it. For her doctoral thesis at Harvard's School of Public Health in 2016, she biked all over the Boston area, measuring and recording sound and also shooting photos and videos at some four hundred locations. She also interviewed locals about their perceptions, memories, and sensitivities related to sound. Bringing together this research, she created an alternative to decibel-based noise maps. She combined her trove of multimedia with layers of neighborhood demographic data to create Boston's first citywide noise report in more than four decades, which she posted online. It included interactive tools like neighborhood maps that linked acoustic measurements to the tally of noise complaints. She also launched an app called NoiseScore, which she hoped could both help people document local noise problems and provide data to one day inform more holistic noise metrics.

For years, Walker readily responded to anybody seeking her help to reinforce their noise struggles with data and the scientific evidence of noise causing harm. By 2021, however, she was most excited about a new initiative in a place where noise is far from the most pressing concern—her hometown, Jackson, Mississippi. While there is plenty of noise in the Jackson area from the usual suspects, such as highways, industry, and freight trains, the city had lately made headlines for one public health catastrophe after another (none of them noise related): devastating floods, skyrocketing homicide rates, and an ongoing water crisis caused by crumbling infrastructure that poisoned tap water with lead and sewage overflow during heavy rains.

More than 80 percent of Jackson's residents are African American, like Walker, and about a quarter of them live below the poverty line. While Walker remained dedicated to fighting noise pollution and was well versed in its harms to health, she wanted her work to empower communities above all else. To that end, she won a grant to fund equipment for measuring noise and for sampling the air and water, to test for toxicity. She planned to provide students and teachers at local high schools and community colleges with background information about these pollutants and the training to gather high-quality data on their exposures. Critically, this would mean that the students and teachers themselves, not Walker or any outside authority, would decide if addressing noise was a top priority.

"I don't like the idea of telling people what they should be concerned about," Walker said. "I'm very passionate about noise. But I'm more interested in making sure the right people in the right places have the tools and capabilities to make a difference in people's lives. If it's with noise, then it's with noise. If not, then not."

Too Loud and Too Quiet in Detroit

The fight against noise pollution is further complicated by the fact that there is no straightforward measure of progress, as there is with chemical toxins. While too much noise undeniably harms the health of a community, too much quiet can sometimes be a symptom of deeper trouble. The silence of vacant lots, empty sidewalks, and boarded-up homes and storefronts are the hallmarks of urban blight. In Detroit, both ends of this spectrum can be found within blocks of each other.

The Southwest Detroit Community Benefits Coalition spent years negotiating with the builders of the new cross-border bridge—a consortium funded by the Canadian government—and with Detroit authorities to win compensation for people directly in the path of the project. In 2017, they struck a deal to carve out $48 million for community investment, including two separate protections from the additional air and noise pollution. Some households would be offered soundproofing via new windows and insulation, plus heating and air-conditioning systems with high-efficiency filtration to block particulates. People living even closer to the future bridge would be offered an entirely new place to live in a less-polluted part of the city; they could choose a renovated home from the thousands of vacant houses owned by the Detroit Land Bank Authority.

Eligibility for these two programs ended 300 feet north of I-75, which left Thomasenia Weston two blocks too far away, and frustrated. From her porch, she could clearly hear the pile driving just to the south, where land had been seized and homes razed to make room for the new highway ramps needed to handle the coming influx of traffic.

Weston gestured toward the banging. "How can they tell me 'You're not impacted by this, you can't hear this, it's not bothering you'?" she asked. "They turned this into a footage issue instead of a fairness issue."

A sliver of the negotiated community compensation had been set aside to continue research into health and pollution exposure throughout the construction of the bridge and after its completion. A handful of air-and-noise monitors remained scattered around Southwest Detroit, dutifully taking measurements. In theory, the findings could bolster the case for expanding the renovation and home-swap programs after the new bridge was in place, although advocates admitted quietly that the chances of this were slim.

Weston, for her part, had pinned her hopes for relief from noise and soot on a proposed truck-routing ordinance, the first ever for Detroit. In 2021, the city hired the planning and engineering firm Giffels Webster to study potential alternative truck routes, and their report made several recommendations for steering trucks onto commercially eligible streets. However, a legal review of the draft ordinance delayed progress. In the spring of 2023, Detroit officials told Weston and her neighbors that no further action would be taken until the completion of a second, more comprehensive study of trucking routes and impacts citywide. A year later, they were still waiting.

Toward the end of our interview, Weston showed me photos taken earlier that summer, when heavy rains made the sewers overflow and basements flood throughout Southwest. When the water receded, it left behind growths of mold and mildew, which worsened her family's asthma and added more cracks and potholes to the street, which was already compromised by previous flooding. The damage to the road's surface made truck traffic even louder. Weston had considered moving but said she couldn't afford it. In addition to her asthma, she struggled with diabetes, arthritis, and atrial fibrillation. Her poor health had recently forced her to trade steady work as a pharmacy technician and medical assistant for running a food cart, which she rolled out at parties and special events.

"My home is not going to sell for anything, between the water issue, the truck traffic issue, and the foundation issue," she said. "So, I feel like I'm stuck."

I later spoke to Rico Razo, director of Bridging Neighborhoods, the city agency administering Detroit's environmental mitigation and home-swap programs. I asked him about the 300-foot boundary that inevitably meant some houses on a block would get new windows, insulation, and cleaner conditioned air, while next-door neighbors' houses got nothing.

"It's strictly a budgetary limit," Razo explained. "If we're talking about actual health impacts from being near a freeway, you really want to push that buffer out to 1,500 feet. But the community benefits were negotiated, and the funds were limited."

At the time of this interview, in the fall of 2021, Bridging Neighborhoods had retrofitted about 140 homes against noise and air pollution, and they expected to complete another couple dozen renovations by the following spring, when the program was slated to end. Home swaps took longer, and Razo said they hoped to have about a hundred complete by the end the fiscal year in June 2022, and about 150 total over the next couple of years, or until the money ran out.

A color-coded map on the Bridging Neighborhoods website displayed the boundaries of community benefits. A green area north of I-75 was designated for home renovations. Immediately south of that, a yellow strip of homes hugged the highway, and their owners had the choice to relocate or have their home retrofitted. According to Razo, most folks in the yellow zone opted to stay put. Below that yellow zone, the entirety of Delray was a blue home-swap-only region, except for a red gash of eminent domain buyouts that had been made to accommodate the bridge and a future customs plaza. The implications of this map were clear: within a few years, yet another community, one already pummeled by decades of environmental abuse, would be completely silenced.

An early settler and veteran of the Mexican-American War had named Delray after Molino del Rey, a town just outside Mexico City where a battle was fought in 1847. Initially an autonomous riverside village, Delray was annexed to Detroit in 1906, and it boomed along with the city's industry. In the 1920s and '30s, loud and vibrant Delray was home to some 24,000 residents— African Americans and immigrants from Poland, Hungary, Mexico, and Ireland, among other places. Many worked in the local factories and patronized the shops, restaurants, and bars lining the neighborhood's main drag, West Jefferson Avenue.

But by the end of World War II, Delray's decline had begun. The downturn accelerated in the 1960s and '70s, as families moved out and more heavy industry moved in; the neighborhood was soon dominated by steel and cement plants, a wastewater treatment facility, a coal-fired power station, and an oil refinery, not to mention the train yards and the massive highway cutting Delray off from the rest of the city. By 2015, when city authorities started forcing buyouts and demolishing houses to make room for the planned bridge and customs plaza, only about 2,000 Delray stalwarts remained.

Under the terms of the home-swap program, the name Delray became synonymous with environmental degradation, with the highest "Delray scores" determining who got dibs on residences elsewhere in the city. A Delray score was based on an accumulation of points that took into account a home's proximity to the new bridge and the number of years its owners had lived in the neighborhood. People earned more points toward their Delray score if they had kids, if they were elderly, or if a doctor could attest to a medical condition making them especially sensitive to air or noise pollution.

By the time of my visit, Delray had shrunk to about three hundred households, and nature was already reclaiming the place. In contrast to the densely packed streets north of the highway, much of the neighborhood resembled a prairie with sidewalks. Entire blocks were nothing but a street sign and a fire hydrant poking defiantly above wild grasses and patches of purple thistle, goldenrod, and Queen Anne's lace. With almost no cars rolling down the cracked streets, no pedestrians, no kids chasing each other around playgrounds, nor any of the usual urban hustle and bustle, it was certainly quiet. In fact, when a quick burst of birdsong rose above the dull traffic rhythms of the still-distant highway, it almost felt peaceful. Nevertheless, the eerie quiet of loss was unmistakable.

Nowhere were the complexities of fighting noise pollution more evident than here. The forces inundating Southwest Detroit with nonstop cacophony had also foisted this desolate stillness on Delray. Noise is rightly understood as pollution, a toxic by-product of modern life that's responsible for widespread, albeit unequal, harm. But the antidote to too much noise will never be as straightforward as it is with other pollutants, for which less is always better and zero is ideal. Ultimately, people want sonic surroundings that reflect a healthy, empowered, and prosperous community, and those goals can't be easily plotted on a decibel chart.

Closer to the river, Delray's vacant lots were backed by black silt fencing, beyond which rose mountainous piles of earth on the edge of the bridge's 165-acre construction site. Just offshore, on Zug Island, the hulk of a steel mill that had idled most operations in 2020 stood like a rusty skeleton against the gray sky. Nearby rose the grim smokestacks of a coal-fired power plant that had belched thousands of tons of sulfur dioxide and nitrogen oxides into Delray's air every year for more than six decades before finally shutting down in 2021.

The construction site was idle that day, and only two massive concrete towers, one decorated with a painted bald eagle and the other with an American flag, hinted at the future presence of the new bridge. The towers were surrounded by cranes, backhoes, and other earth-moving equipment, and two towers mirrored them on the Canadian side of the river. Nothing lay between them yet, except empty air and a massive silence. It felt as though the place had just taken a very deep breath and was awaiting the din to come.

5

SENSORY SMOG

NATURE IS LISTENING

We typically understand noise to be a "people problem," created by us, perceived by us, and harming us in the form of hearing loss, distraction, and health ailments brought on by chronic stress and broken sleep. But noise pollution is nature's problem too. The spread of our mechanized din, alongside the glow of artificial light, into wild areas has brewed up what conservationists have dubbed a "sensory smog," hindering animals as they try to communicate, sense the presence of predators, hunt for food, and simply find their way.

Few people understand nature's growing noise problem better than Kurt Fristrup, a bio-acoustics expert who recently retired from the Natural Sounds and Night Skies Division of the National Park Service (NPS). One sweltering June afternoon, Fristrup led me on a hike in Colorado's Lory State Park. Tall and wiry in his mid-sixties, he climbed the trail quickly until we reached Arthur's Rock, a bald, sun-drenched outcrop overlooking the slate-blue stillness of Horsetooth Reservoir and the city of Fort Collins, where the NPS Natural Sounds Division is headquartered.

A breeze whipped across our exposed perch, but I could still hear the rapid-fire trills of a western tanager, one of several birds Fristrup had identified during our hike, along with the zippy alarm call of a yellow-breasted chat and the clear, dipping notes of a canyon wren. At least, I thought it was a tanager. Fristrup had described the call as similar to a robin's, but faster and raspier. I sipped from my water bottle and peered up into the thin needles of the Douglas firs, searching for a telltale flash of tanager yellow and thinking of something else Fristrup had said during our ascent.

"I've come to believe," he'd said, "that when we mask nature's sounds with more and more noise, we haven't simply lost opportunities to hear those sounds; we've conditioned people not to listen."

He was talking about more than birdsong. Specifically, humans have adapted to the noise in our lives at least partly by tuning it out, by not listening. When we need quiet, we grab a pair of noise-canceling headphones, silence our smartphones, and shut out the world for a while. Animals don't have that luxury. Wildlife is always listening, or at least trying to, despite the insidious encroachment of human noise into their domains from roads, sprawl, and several other sources that Fristrup and his NPS colleagues had been tracking for years. Fending off this sensory smog will take a resurgence of listening on many levels.

Umwelts: How Nature Listens

As we gaze across an expanse of open water, it's easy to imagine a vast, soundless world beneath the waves. The notion of the silent watery deep has a certain poetry befitting the ocean's enduring mystery, and it makes intuitive sense to those of us whose time submerged is typically limited to single breaths within steps of shore. The instant we dive below, the noise of everyday life vanishes, and we can revel in serene weightlessness for a few moments until our lungs force us back to the surface.

But in fact, the aquatic world is filled with symphonies of sound. The lash of rain and wind on the surface echoes below, while geothermal vents gurgle and hiss furiously on the seafloor. Fish chirp and grunt to each other, whales sing their slow, eerie hellos, and legions of snapping shrimp emit a constant crackle that's powerful enough at close range to stun their tiny prey.

Discovering how the inhabitants of this watery world perceive and use sound to find food, attract mates, and avoid predators is the first step to understanding how noise pollution threatens all these critical behaviors. A creature won't be the least bit bothered by sounds in frequencies outside its perceptual range, for instance, no matter how loud and persistent the noise is, but it may be extremely vulnerable to noises that directly interfere with sounds it relies on for its essential functions. In short, we can't reckon with the danger noise poses to wildlife without understanding their unique sensory worlds—dubbed "umwelts" by the early-twentieth-century Baltic German zoologist Jakob Johann von Uexküll.

Hearing is especially important for many animals living in the ocean's expanse, where sound travels farther and faster than it does through the air and where light to see by is at a premium. At certain depths, for example, whale vocalizations can be heard across thousands of kilometers, while both baby reef fish and coral larvae begin life in the open ocean and find their reef homes partly by listening for them.

Nevertheless, scientists have only recently started to investigate the acoustic umwelts of undersea creatures, which differ greatly from our own. Indeed, many aquatic species lack functional external ears. Fish, for example, have microscopic hairlike cilia beneath their skin to sense the motion of water particles jiggling back and forth due to sound-wave energy, and they make their own sounds, despite having no vocal cords, by rubbing together their bones and teeth—an action known as stridulation—or by using specialized muscles to drum on the gas-filled swim bladders that keep them buoyant.

Among the world's leading experts on undersea umwelts is Aran Mooney, a sensory ecologist at the Woods Hole Oceanographic Institute on Cape Cod. He has spent a career piecing together the sonic realities of marine species. On a recent mid-September morning, Mooney was at work on a weathered pier in Woods Hole. Sun lingered on the working waterfront, but a stiff breeze stirred up a light-green chop where the harbor opened into Vineyard Sound, heralding a late-summer storm that was creeping up the Atlantic coast.

In his mid-forties and clad in a T-shirt, shorts, sporty sunglasses, and a mesh-backed baseball cap tugged over a fading buzz of brown hair, Mooney and a few lab mates were hurrying to complete an experiment before the weather turned. They wanted to find out how squid react to the hammer blows of pile driving, which can top 200 underwater decibels (equivalent to about 138 decibels through the air)* and can be detected more than 6 miles away.

---

* Recall that decibels are a relative, not an absolute, measure of sound intensity, and water is a denser medium than air, so the same sound intensity equals more decibels in water than it does in air. The rule of thumb is to subtract 61.5 decibels from the underwater total to determine the intensity of the same sound energy through the air.

Mooney has a long history of working with squid, dating back to 2010 when he led the first-ever studies of squid hearing. That investigation grew out of earlier research on how noise impacts dolphins, some species of which hunt squid by means of echolocation. Bursts of ultrasonic clicks emitted by the bulbous "melons" on a dolphin's forehead bounce off prey and return to vibrate the dolphin's lower jawbone, which conducts sound signals directly to the middle ear. Mooney wanted to know if squid could hear an approaching dolphin as it tracked them and sense the clicks indicating their impending demise. First, though, he had to answer a more basic question: do squid hear anything at all?

The answer was yes. Squid could detect sound, but only at low frequencies topping out at around 500 hertz (about one octave above middle C on a piano). Mooney discovered that squid and other cephalopods hear via two roundish, fluid-filled sensory receptors inside their head called statocysts. Within each of these receptors is a tiny calcium crystal that sloshes around when buffeted by sound-jostled water and brushes up against a carpet of sensory cilia. Since this initial research on squid, Mooney and his colleagues have investigated the acoustic umwelts of many more marine species, from seabirds to turtles to coral polyps.

Back on the dock, the Woods Hole researchers netted one squid at a time from bathtub-sized holding tanks and fitted them with tiny motion sensors before transferring them into underwater cages set at various distances from a hulking steel pile poking out of the water. A massive hydraulic hammer hung from a crane in an adjacent parking lot, waiting to pound the pile several feet deeper into the harbor mud.

Experiments by Mooney and others have shown that underwater loudness does matter, although it's far from the only threat that noise poses to aquatic creatures. They've found that intense underwater sound can trigger temporary hearing loss in many marine species. While sensory cells regenerate in the hearing systems of fish, amphibians, and birds, no hair cell regrowth occurs in marine mammals such as whales and dolphins. That said, the Woods Hole researchers on the pier that day were not investigating squid hearing loss. Their pile-driving study had other aims.

For starters, the local squid fisherman who donated the animals for this experiment worried that pile driving during the construction of a proposed offshore wind farm would scare away the source of their livelihood. Squid fishing is big business, generating nearly a quarter billion dollars annually for the New England and the mid-Atlantic economies, according to a 2020 analysis by the Science Center for Marine Fisheries. But Mooney had a wider range of concerns. Squid are a keystone species at the center of a food web in which everything either eats squid or is eaten by squid. So, if noise causes squid to flee, and their population in a given stretch of ocean plummets, then the impacts of that loss will cascade throughout the local ecosystem.

Even if pile driving didn't cause squid to flee, Mooney explained, the noise it creates could still harm squid by disrupting essential behaviors such as feeding or escaping from predators. In 2019, he and fellow researchers played recordings of pile driving through underwater speakers in tanks filled with squid and the tiny fish they hunt. In these noisy waters, the squid were less effective at catching and gobbling down fish. At the same time, their initial startle response to the loudness—shooting off plumes of ink and jetting off to escape—dissipated after the first few rounds of noise. Becoming habituated to disturbances in this way could prove problematic when real predators draw near.

More broadly, research by Mooney and others suggests that noise can harm animals simply by robbing them of energy they can't afford to lose—exacting a caloric cost from a limited budget of energy. Animals need to woo mates, fight off rivals, care for their young, flee predators, and migrate, and a noise-saturated environment makes all those activities more taxing. In the pile-driving study, for example, the tiny motion detectors might show that squid moved their fins faster when exposed to more noise, wasting precious energy on hypervigilance. If noise also made squid worse at hunting, as the lab research indicated it did, then that would be a caloric double whammy.

"These guys don't have a lot of fat storage," Mooney said of the squid. "If they burn through their energy, then there's not a lot of reserve for them."

The Everywhere Problem

The ubiquity of noise in our increasingly crowded oceans matters as much as, if not more than, its decibel count. Yet chronic underwater noise pollution is often overlooked because its insidious harms remain hidden beneath the waves and because extreme loudness is more dramatic, with graphic evidence of the danger it poses to undersea creatures sometimes washing up on shore, as it did in the Bahamas on March 15 and 16, 2000.

Over the course of those two days, the US Navy was engaged in antisubmarine training, using sonar pulses of up to 230 underwater decibels. This noise hounded sixteen whales into the shallows of the Bahamas, where six of the massive animals slowly died from overheating. The mass stranding lured TV news cameras, sparked congressional hearings, and helped launch the career of the marine biologist Brandon Southall.

At the time, Southall was completing his doctorate at the University of California, Santa Cruz, on Monterey Bay, where he had conducted some of the earliest studies on how high-intensity noise damages the hearing of marine mammals. After environmental groups filed multiple federal lawsuits against the navy to restrict sonar use, Southall was called as an expert witness in cases that made nationwide headlines. He then joined the National Oceanographic and Atmospheric Administration (NOAA) and moved to Washington, DC, where he became the agency's director of ocean acoustics and briefed President George W. Bush's scientific advisers about sonar's impact on wildlife.

Southall left NOAA in 2009 and returned to Monterey Bay to launch an independent research consultancy called Southall Environmental Associates (SEA, Inc.)* with his wife and fellow scientist, Kristin Southall. More than a decade later, he was still doing sonar research, but he worried that the spectacle of beached whales, federal lawsuits, and congressional hearings had overshadowed the much bigger underwater noise hazard emanating from tens of thousands of tankers, cargo vessels, and other commercial ships churning through the oceans every day.

Most vessel noise is caused by cavitation (the formation of air bubbles) around the propeller. These bubbles rapidly build up and implode due to differences in pressure around the spinning blades. The resulting sound can range in loudness from

---

* He maintains research and teaching affiliations with the University of California, Santa Cruz, and Duke University.

about 170 to 190 underwater decibels within a few meters of the ship and carry for miles before fading away. Sonar, pile driving, and the air-gun blasts of seabed explorations for oil and gas are all louder, but they are also localized and episodic, Southall explained. "Vessels are everywhere, and their noise never stops. It never shuts off."

And the problem is getting worse. A 2012 analysis in the journal *Nature* found that the ambient sound levels of the deep ocean had been rising about 3.3 decibels per decade since 1950, thanks largely to the low-frequency signature of heavy ships (under 1,000 hertz), which overlaps the sweet spot of vocalizations for many baleen whales. That accumulation of noise amounted to nearly 20 extra decibels of constant *background* sound, an extraordinarily rapid shrinking of the whales' sensory world over a time period that's well within the lifespan of many whale species.

Marine biologists have observed that whales respond to this incessant sonic assault much like people do in a noisy bar. They raise their voices, stay closer together, and simplify what they say. In one study, scientists attached suction-cup acoustic recorders, known as D-tags, to North Atlantic right whales to track their social "up calls," which are brief, rising whoops the whales use to stay in touch with each other. The researchers found that whales tried to keep a 10–12 decibel advantage over the background sound, which meant their up calls ranged between 117 decibels in relative quiet to 150 decibels—essentially, whale shouting—when things got noisy. Every one of these extra-loud vocalizations saps energy. Meanwhile, other whales just give up. In 2018, a team of Japanese researchers found that humpback whales simply stopped calling when ships passed near them, and the whales remained silent for up to half an hour.

Vessel noise also makes whales less efficient at foraging for food. Humpback whales, for instance, hunt little silver fish called sand lances by diving deep and executing a series of acrobatic rolls to corral the schooling fish for easier gulping. But when a passing ship makes things noisy, humpbacks become less intense hunters, diving slower and making fewer turns, according to an acoustic-tag study in the Stellwagen Bank National Marine Sanctuary, a biologically rich area of the Atlantic Ocean about 25 miles east of Boston. Because of Stellwagen's biological abundance and proximity to a major port, the area is abuzz with vessel traffic, including fishermen, whale watchers, and commercial shipping. In fact, vessel noise was so constant during the study that researchers struggled to find relatively quiet periods to make comparisons and so had to analyze data gathered at night, when ships passed through more intermittently.

An ever-shrinking umwelt challenges any animal, but it can be especially threatening to an endangered species like the North Atlantic right whale. Commercial hunting throughout the nineteenth century pushed these whales to the brink of extinction, but a hunting ban enacted in 1935 allowed for an uneven recovery. Numbers of right whales crept upward in subsequent decades, reaching an estimated 500 animals in 2010 before suddenly declining again to about 360 whales as of 2024.

While scientists remain puzzled by the right whale's recent setback, they have identified a mix of potential culprits. First, right whales stay close to shore during their seasonal migrations up and down America's eastern seaboard, which increases the risk of being hit by ships and entangled in fishing gear. In addition, female right whales are having fewer calves, and the recovery time for females between births has doubled, from three to five years to six to ten. The slowdowns in calving are likely tied

to declines in nutrition. Foraging whales must gobble down billions of zooplankton every day to sustain their 70-ton bodies,
and as climate change shifts the availability of this prey, the
whales must spend more time and energy finding enough to eat.

Pervasive underwater noise compounds all these challenges—
from migrating to mating to foraging. Leila Hatch, a marine
ecologist with NOAA's Office of National Marine Sanctuaries,
determined that right whales in the crowded waters off Massachusetts have lost about two-thirds of their acoustic range
compared with historical periods when the ocean was 10
decibels quieter.

"In a quieter ocean, these animals could hear a mate or family member call from twenty kilometers away, but that's now
reduced to a kilometer or two," Hatch explained. "Noise just
adds to the fabric of what's making it harder for these whales
to get by."

Another way to understand noise's impacts on animals is to
observe them during increasingly rare moments of quiet. For
example, in the strange hush that descended during the first
year of the Covid pandemic, whales were among several animals to happily reclaim their sonic space. Massive cruise ships
stopped arriving in the jade-green waters of Alaska's Glacier
Bay National Park in the summer of 2020, cutting the median
daily levels of low-frequency sound by two-thirds, compared to
the previous summer. Marine scientists from Cornell University and the National Park Service tracked the response from
local humpback whales. During these months, the whales' calls
grew more varied and carried for nearly a mile and a half, compared to only 650 feet in pre-pandemic waters. The pods of
humpbacks also spread out, extending their feeding grounds
while emitting fewer "whup" contact calls, similar to the pro-

portion of whup calls from humpback recordings made forty-five years earlier.

Shipping soon returned to pre-pandemic levels, however, and the underwater din is poised to spread into the Arctic as polar ice melts and the few remnants of relatively undisturbed ocean are opened to commerce and the exploration and extraction of natural resources. Canadian researchers reported that between 2013 and 2019 the number of ships in the Arctic jumped by 25 percent and covered 75 percent more distance, doubling the average noise levels in those waters. In 2021, more than two dozen scientists from around the world raised the alarm about underwater noise, coauthoring an article in *Science* titled "The Soundscape of the Anthropocene Ocean." Sifting through more than 10,000 papers, the authors found evidence of sonically degraded habitats for everything from invertebrates to fish to marine mammals. They labeled underwater noise "a pollutant that cannot be ignored."

A Listening Science

Humans have cultivated much stronger bonds with the sounds of terrestrial wilderness than they have with underwater soundscapes, simply because we spend a lot more time listening to them. In fact, the National Park Service has had a mandate since its founding in 1916 to preserve the parks' natural resources, including their soundscapes, and yet it was only in 2000 that the park service established the Natural Sounds Program* and truly started listening to them.

---

* In 2011, NPS expanded and upgraded the program, calling it the Natural Sounds and Night Skies Division.

That year, Congress told the FAA and NPS to jointly craft air-tour management plans for the increasingly crowded and noisy skies over the Grand Canyon and other popular parks.* But before NPS could devise new rules to stave off air-tour noise, they needed baseline recordings of what they sought to protect. So the agency sent young staff and interns trekking into the wilderness. They lugged microphones, tripods, cables, ruggedized laptops, and sound meters nestled in hard plastic cases and carried enough power to operate all the equipment for twenty-five straight days and nights of recording.

A quarter century later, the NPS Natural Sounds Division boasts its own fabrication lab to customize field-ready recording kits, with printed circuit boards and a computer-guided router for fashioning whatever is needed out of lightweight waterproof plastic. They've amassed thousands of field recordings from the parks, and one fact they have established beyond doubt is that human noise is everywhere in nature, even in the most protected wilderness areas.

In her office in Fort Collins, the NPS acoustic research specialist Emma Brown played me a sound file on her computer recorded from southern Colorado's Great Sand Dunes National Park. The room was filled with the nocturnal yips and howls of a coyote pack and an autumn elk rut—when male elks emit loud, high-pitched bugle calls edged with deeper, menacing groans to find mates and deter rivals. At one point, we even heard the thud of elk footfalls and the clatter of clashing antlers.

* This move was catalyzed by a collision between sightseeing aircraft over the Grand Canyon in 1986, which killed twenty-five people and spurred passage of the 1987 National Parks Overflights Act, mandating that the impacts of overflights be studied throughout the park system.

For each sonic event, Brown quickly pointed out the corresponding acoustic signature on a spectrogram scrolling across her computer. Ragged peaks and valleys followed changes in frequency over time, color-coded for intensity: ultraquiet blue and violet below 25 decibels, then more energetic orange and yellow, and all the way to white, the loudest sounds, above 70 decibels. Common human-made noises were easy to spot due to their regularity, such as the yellow curves of overhead jets. In other recordings, passing snowmobiles in Yellowstone and fan boats in the Everglades grew into thick white lines that briefly obliterated every other sound.

Of course, all this data must be sorted and analyzed. For now, the most accurate way to dissect a soundscape is with trained human listeners, although artificial intelligence systems are catching up. Machine learning can recognize the vocalizations of specific species and tag each occurrence across many hours of recording. For instance, Brown and her team make extensive use of BirdNET, an app that uses artificial intelligence to identify the species of birds heard in real-world recordings, which was developed by the Cornell University Lab of Ornithology. Still, a complete soundscape annotation, including the many human-noise disruptions, is still too much for a robo-listener. In 2013, NPS partnered with faculty and students at Colorado State University in Fort Collins to help manually annotate the crush of field recordings in the university's Listening Lab.*

In 2017, Emma Brown, Kurt Fristrup, and other researchers from NPS and the Listening Lab analyzed nearly 47,000 hours of recordings amassed from about five hundred protected areas. In one-third of the sites, the median loudness of human-made

---

* In 2024, the parks department entered a second Listening Lab partnership with Penn State University.

noise—from overflights, road traffic, logging, and other sources—
was 10 decibels greater (that is, ten times louder) than the background sounds of nature. At more than three-quarters of the
sites, human sound had at least a 3-decibel edge (roughly two
times louder). Nevertheless, the bulk of these sonic intrusions
topped out between 60 and 70 decibels.

As with underwater environments, the most harmful sounds
in terrestrial ecosystems are not necessarily the loudest but
rather the most persistent. About a decade ago, for instance, a
team of biologists at Boise State University showed how chronic
noise degrades bird habitats in studies featuring a "phantom
road." Along a half-kilometer stretch of a wooded ridge in southern Idaho, the researchers positioned fifteen pairs of loudspeak-
ers, facing in opposite directions, that played traffic sounds at a
modest 55 to 60 decibels by day and then went quiet at night, to
mimic real traffic patterns. No bird's hearing was put at risk by
this modest decibel disturbance. At the same time, the phantom
road produced no exhaust fumes, no bright lights, and no collision risks. Chronic noise was the only pollutant.

The first study took place over several weeks in early autumn,
a time when several migratory bird species settle along the ridge
to rest and refuel before continuing their arduous journey south
for the winter. The scientists alternated four days of road noise
with four days of silence while keeping count of the birds. Overall, bird numbers dropped by 25 percent, and some species completely avoided the area on days with road noise.

A few years later, in a second phantom-road study, the
researchers examined birds that had toughed it out on the
ridge despite the noise. These animals were in rough shape
compared to birds examined during quiet times, averaging
a full standard deviation lower in relative body mass, which

was bad news for birds that still had a long migratory flight ahead of them.

Why would traffic noise make birds skinny and bedraggled? In follow-up lab work, the researchers watched captive birds from a species that frequented the ridge, the white crowned sparrow, as the birds foraged for food in quiet or during road-noise playback. During periods of noise, the sparrows spent 30 percent less time eating and 21 percent more time with their heads up—a sign of vigilance. As their ability to hear approaching threats diminished, the birds instinctively looked around rather than gathering the fuel they would need to complete their migration.

Because nothing in nature changes in isolation, the impacts of noise pollution can ripple through an ecosystem. For example, biologists at California Polytechnic State University spent several years studying the effects of noise on the flora and fauna of Rattlesnake Canyon in northwest New Mexico, an area dominated by woodlands of pinyon pine and Utah juniper interspersed with sagebrush, which also hosts a plethora of natural gas wells. Some wells are coupled with noisy compressor stations that run day and night to maintain stable pressure in the gas lines, while other wells have no compressors and are comparatively quiet.

The scientists discovered that noisy wells drove away two birds that are key seed distributors for the canyon's primary tree species—the Woodhouse scrub jay, which buries pinyon pine seeds to store them for winter, and mountain bluebirds, which eat juniper berries and spread the seeds in their droppings. In 2007, the scientists established 115 survey plots around both noisy and quiet wells and tracked the tree seedlings that grew in the plots over the next twelve years. Plots near wells with noisy compressors had 75 percent fewer pinyon pine seedlings and

86 percent fewer juniper seedlings than the quieter plots did. Noise had hit these areas like an invisible wildfire.

Slowing the Spread of Sensory Smog

Reducing our noise impacts in nature will be neither simple nor uncontroversial. People are part of ecosystems, after all. Even the natural wonders of America's national parks were heavily peopled by Native Americans before they were pushed off their lands.

The primary role of the Natural Sounds Division at NPS is to provide technical assistance for noise-related issues that arise inside park boundaries. While parks have enacted several highly contested measures to reduce noise, such as restricting snowmobile access,* Brown and her team try to take an awareness-first approach. They assume that most visitors are motivated to protect the parks but just don't realize the extent to which their noise can harm the natural wonders they've come to experience.

For instance, every August, hundreds of thousands of motorcycle riders gather for a nine-day rally in Sturgis, South Dakota, and their itinerary often includes a ride to Devils Tower National Monument, a massive butte in the Black Hills about 80 miles to the west. Research by NPS and Colorado State University had shown that noise from thousands of motorcycles was disturbing bats, deer, and a local community of prairie dogs popular with tourists, who stop by the roadside to watch their busy antics.

---

* Many national parks now operate with "noise budgets" meant to allow management some flexibility and to encourage the adoption of quieter technologies.

So, a few summers ago, Brown stationed herself in a tent at a parking lot near Devils Tower, handing out pamphlets to bikers that included information on the impact of noise on wildlife and requested that bikers avoid excessive acceleration, engine revving, and honking. "I had some great conversations with motorcyclists who came up and told me that natural sounds were a really important part of their visit as well," Brown told me. "I don't want to paint a completely rosy picture of it, but education has to be the first line of defense for managing these issues. Noise is not something that the average park visitor really thinks about."

Meanwhile, reducing underwater noise may be an even greater challenge. The closest thing to an overarching authority on the high seas is the International Maritime Organization (IMO), a United Nations agency that sets standards for shipping safety, security, and environmental stewardship. Quieter ships have been on the IMO agenda since 2008, with negotiations continuing among ecologists, acousticians, shipping industry representatives, port authorities, and officials from the handful of "flag state" nations where most ships are registered. These negotiations have focused on consensus building, which has kept the discussions amicable, with no parties storming out, threatening lawsuits, or badmouthing one another in the press. This positive approach has also slowed progress to a crawl.

Technically, we know how to shush a ship. Research vessels designed for enhanced quiet are equipped with precision-engineered propellers, which churn up less cavitation. These changes to propulsion systems can be combined with more streamlined hulls, drag-reducing coatings, and sound absorption in engine rooms, which can together achieve a drop of tens of decibels compared to typical ship noise. Ships with these upgrades are not only quieter but also more efficient, which saves money on fuel. More efficient ships will also help shipping companies meet targets for reducing greenhouse gas emissions. In 2017, for instance, the Dutch shipping giant Maersk plowed about $100 million into upgrading the hulls and propulsion systems of eleven cargo ships to save on fuel and reduce emissions. In subsequent testing, these retrofitted ships were also 6–8 decibels quieter.

However, technical know-how is not enough. Ships are expensive long-term investments, and because vessel noise is an aggregate problem rather than the fault of a few ultra-loud transgressors, any solution must be implemented across the industry. In 2014, the IMO finally issued voluntary guidelines for making ships quieter, and then they waited a few years to see if shipbuilders would comply with them. Most did not. That was no surprise to Kathy Metcalf, president and CEO of the Chamber of Shipping of America, who has represented the interests of shipping companies at the IMO noise negotiations.

"Unless everyone's doing it, no one shipyard is going to spend another million bucks on the propeller and hull refinement needed to build a quieter ship. It would put them at a competitive disadvantage," Metcalf said. In 2023, the IMO updated its voluntary guidelines to emphasize that ships operating in Artic waters near sound-sensitive species need to be quieter, but the organization did not move to make any noise reductions mandatory.

Still, even without industrywide vessel redesign, ships can reduce cavitation and thereby run more efficiently, with less noise—simply by slowing down.* For years, shipping companies have experimented with "slow steaming," meaning they reduce ships' speeds during mid-ocean transit to save money on fuel. Also, speed limits already exist in some coastal waters to reduce the risk of collisions between ships and whales, and the voluntary slowdown zones that a handful of ports have implemented in recent years to promote quiet have enjoyed impressive compliance. For example, since 2017, the port of Vancouver has persuaded about 90 percent of the large ships entering Puget Sound to slow down and shift routes seasonally to avoid disturbing endangered orcas and whales. Noise levels in high-traffic entrances to the sound have dropped by 50 percent during the slowdowns, and other ports, including Seattle's, have since instituted their own slowdown zones.

The Vancouver port authorities made reasonable accommodations to the shipping companies by compensating them for the extra hours that vessel pilots spent navigating the straits, and they took the different types of ships into account as they determined how much they should slow down. The guidelines were also carefully crafted within a broader initiative to monitor underwater noise and visually track orcas and whales to warn ships of their presence.

Indeed, the vastness of the oceans and the immense variety of aquatic umwelts argue against one-size-fits-all solutions. A more tailored and adaptable approach to reducing underwater noise would focus on protecting the soundscapes of critical areas such as coral reefs, spawning grounds, and migration

---

* There are exceptions. Engines are designed to run within an optimal range of speeds, below which they are at greater risk of fouling, corrosion, and other issues that can impair efficiency and increase noise.

corridors, and that degree of specificity will require a lot more long-term listening.

The Call of the Wild: What's in It for Us?

While a lot of research on sensory smog is rightly focused on wildlife, nature is not alone in benefiting from the protection of its soundscapes. Every year, hundreds of millions of national park visitors generate billions of dollars in economic activity, and surveys find that listening to nature is a big part of why people make the trip. Our encounters with wilderness are immersive multisensory experiences, and one feature of a visit can complement another—or not. In one study, for example, people listening to background nature sounds while looking at photographs of scenic overlooks in Yellowstone, Glacier, and Denali National Parks rated them about one-third less scenic if the noise of a snowmobile, motorcycle, or propeller plane intruded on the soundtrack of birdsong and other wilderness sounds.

For most of us, forays into pristine wilderness are infrequent, but nature's sounds still matter to us when they're competing with the din of urban life—perhaps even more so. While the sound of traffic and other such noise ramps up stress and distraction, there's mounting evidence that listening to natural sounds may be an antidote, relieving stress and restoring attention. In a 2022 study, for example, environmental neuroscientists at Berlin's Max Planck Institute found that six minutes of recorded birdsong reduced measures of depression and anxiety in listeners, while people subjected to recorded traffic noise came to feel *more* depressed. Neuroscientists at Kings College in London found similar mood boosts in response to birdsong among nearly 1,300 people over two weeks; they made regular mental-health check-ins via a smartphone app called Urban Mind.

Yet it's not so simple. As we'll explore later, our reactions to nature are both steeped in evolution—à la the biophilia hypothesis popularized by the eminent biologist E. O. Wilson—and tempered by personal histories and cultural memory. For many people, walking a secluded forest path and hearing nothing but birdsong, critter chatter, and wind-rustled branches is the epitome of tranquility. But for others, the same experience might be disorienting or even hint at lurking danger. Whatever the case, most of us rarely find ourselves in the presence of unadulterated natural soundscapes without some amount of human sound intrusion, whether it comes from a nearby road, a jet passing overhead, or simply our fellow visitors.

In 2021, Kurt Fristrup and other bio-acoustics experts weighed the effects from eighteen studies and found that listening to nature improved health and happiness measures by 184 percent and lowered levels of stress and annoyance by 28 percent, on average. Separately, they analyzed recordings from 221 sites across sixty-eight national parks, including many located in or near cities, and tabulated the occurrences of natural sounds and human sounds, such as those made by cars, airplanes, and people talking.

The results were sobering. Even in these protected areas, undisturbed natural soundscapes were rare. Only 11 percent of the sites were free of human sounds more than 25 percent of the time, and unsurprisingly, the most pristine spots were far away from where people live and were difficult to access. At the same time, nature sounds remained abundant even in the most-visited parks. Among the five urban parks analyzed, birdsong and other nature sounds were audible about 60 percent of the time.

"This is where things get really interesting," Fristrup said. "If we can manage noise in these areas, what should be our priorities?"He thinks the next frontier of research should go beyond isolating natural and human sounds and attempt to find out "how much and what kinds of noise can mix into a natural soundscape before it loses its restorative benefits."

Encouragingly, sometimes all it takes to reduce noise is to remind people that there's something to be gained from listening, as the managers of Muir Woods National Monument discovered. Located north of San Francisco, Muir Woods is a lush forest of towering old-growth redwood trees that can be explored by means of walking trails along Redwood Creek. The woods' proximity to the city makes the place a popular destination for tourists and school field trips. In the busy summer season, up to 10,000 people a day pile into the woods, and big crowds bring big noise.

At first, park managers mainly worried about how the hubbub would disturb the forest's endangered spotted owls. But soon they started wondering about how noise might impact other creatures—the human visitors. How did noise affect their experience of the woods? While the forest's well-traveled paths are hardly backcountry wilderness, walking beneath massive old-growth redwoods nevertheless instills a sense of reverence and awe. What acoustic experience did people expect from this unique place? Was there a level of natural quiet that they would seek out and self-enforce?

To find out, park managers partnered with researchers from NPS, Colorado State University, and the University of Vermont to layer different amounts of human noise recorded at Muir Woods onto a baseline of the forest's natural sounds. Hundreds of visitors listened to 30-second selections of these mixes and rated them on a 9-point scale of acceptability, from 4 (very acceptable) to −4 (very unacceptable). The study revealed that the threshold for acceptable human noise (zero on the rating scale) was 37 decibels, a level that was frequently exceeded in the woods by up to 10 decibels during busy periods.

The results suggested that people thought the woods were a bit too noisy and that a little nudge on behalf of quiet might go a long way. So, on randomly selected days, researchers posted signs at the entrance to an area called Cathedral Grove, containing some of the oldest and tallest trees in the woods, to declare it a "Quiet Zone." They asked visitors to hush themselves, silence their cell phones, and "enjoy nature's sounds." On days when the signs went up, the levels of ambient sound in Cathedral Grove dropped by 3 decibels, and 98 percent of visitors surveyed approved of maintaining this quiet zone.*

A follow-up study in 2020 revealed that the intervention had triggered a virtuous listening cycle that benefits visitors and birds alike. This study included bird counts on days with and without "quiet, please" signs throughout the woods. Not only was birdsong easier to hear on the quieter days, but the reduction in human noise also lured more birds to settle in the trailside trees and add their voices to the chorus, which further enriched the experience of those who kept quiet and listened.

Whether our aim is to improve our wilderness experiences, protect certain species, restore habitats, soothe our psyches, or all of the above, there are no fail-safe formulas for rolling back the sensory smog we've unleashed on the natural world. Nevertheless, we should begin. If we wait until we have perfect information, we will wait forever. Compared to how long we've been

---

* Cathedral Grove has since been permanently designated as a quiet zone.

polluting nature's soundscapes, we have barely begun to tally our noise impact on umwelts beyond our own. It will take a lot more listening to find solutions. The challenge is immense, but Kurt Fristrup offered a note of optimism before we left Arthur's Rock at Lory State Park and began our descent:

"In probably every talk I've given about noise, the dominant theme and my closing remarks are that noise is unique as an environmental pollutant because it's so ephemeral," he told me. Decibels don't linger in the atmosphere for centuries, as greenhouse gases do. Noise neither accumulates in the soil nor floats along in the water like microplastics waiting to be eaten by fish, and eventually by us.

"Once you reduce noise by treating its source, it's gone, and the recovery can begin right away."

PART 2

A Better-Sounding World

6

BEYOND NOISE

A WORLD OF UNBOUNDED SOUND

In the spring of 1930, a truck roamed New York City's streets on a mission to defeat the urban din. Dispatched by the city's health commissioner, the truck was crammed with sound meters, vacuum-tube amplifiers, microphones, and other acoustic gadgetry festooned with dials, gauges, and glowing indicator lights. The noise fighters were serious men in dark suits, fedoras, and trench coats who covered some 500 miles, stopping at 138 noise hot spots and taking more than 10,000 measurements of sound. Bell Labs supplied all the din-slayers' equipment as well as their most powerful weapon—the recently devised decibel.

The mobile lab was far from the first effort to combat urban noise. Authorities in ancient Rome sought to protect citizens' sleep by prohibiting nighttime chariot driving. Middle-class intellectuals in Victorian England led anti-noise campaigns as they struggled to think and write in clamorous cities that were rapidly filling with factories (and immigrant laborers). New York City itself already had a prominent history of noise fighters, including the extraordinary Julia Barnett Rice—musician, medical doctor, mother of six—who founded the Society for the Suppression of Unnecessary Noise in 1906. What began as Rice's personal war against the Hudson River tugboat whistles that were disrupting domestic life in her Upper West Side mansion burgeoned into a citywide crusade for quiet around schools and hospitals.

Nevertheless, New York's noise truck, and the landmark "City Noise" report that would be based on its work, was the harbinger of modern noise control, and this was because of the decibel's civic debut. Armed with this new metric, noise fighters could take the scientific measure of their enemy, detect it in no uncertain terms, and target it for elimination. The city passed its first noise code in 1936, and in the decades that followed, the decibel became the sine qua non of noise. Wherever sound was raised as a concern, from the development of nuisance regulations to the rulings of licensing boards, the decisions focused on this single measure, albeit sliced and diced in various ways. Decibels soon informed the answer to almost every question related to noise. Does that sound qualify as noise? Is it potentially harmful? Is it a sound that architects, engineers, or city planners need to take into account? In every case, the decibel meter would tell the tale.

Yet despite a century of chasing down decibels, our world is arguably noisier than ever. Battling unwanted sound can often seem as futile as fighting the tides because so much of what we enjoy and rely on in life makes sounds that may bother someone else. We demand the next-day delivery enabled by all those trucks rumbling through residential streets. We crave more flights to popular destinations, but we hate the roar of airplanes crowding the skies overhead. And of course, we drive and drive and drive.

Meanwhile, traditional noise control is inherently reactive, confronting sounds only after they become entrenched in our lives and churn out enough decibels to trigger the usual counter-measures. Try as we might, we haven't been able to decibel-hunt our way out of this fix. Where does this leave us?

Shifting Soundscapes

Recognizing that noise is about a lot more than loudness leads, in turn, to the realization that pursuing fewer decibels is not the only solution to our sonic problems. Rather than focusing narrowly on decibel thresholds, above which sound is problematic and below which it can be largely ignored, we should try to foster sound environments that support our endeavors. While traditional noise control remains necessary, it's not sufficient for reaching this larger goal.

We should take a more proactive and holistic approach to shaping our sonic world—a growing number of sound and noise experts refer to this as the "soundscape approach." Its roots date back half a century to the late Canadian composer R. Murray Schafer, a faculty member at Simon Fraser University in Vancouver. Schafer founded the field of acoustic ecology, studying soundscapes as emergent compositions of human and natural sounds unique to specific places.* In the 1970s, he led teams of graduate students and young composers in a massive undertak-

---

* Schafer didn't coin the concept of soundscape, which first appeared in a 1969 article by Michael Southworth, an emeritus professor of urban design and planning at the University of California, Berkeley. Southworth wrote *The Sonic Environment of Cities* as a doctoral student at MIT before moving on to other topics in urban life. Schafer's writing and teaching on soundscapes popularized and expanded the concept over subsequent decades.

ing to record and analyze the world's soundscapes. While Schafer and his colleagues gathered basic acoustical data, like decibels
and frequencies, they considered perception and context to be
far more important for understanding a city's soundscapes.

Fanning out across Vancouver in their first major soundscape
study, they interviewed residents as "earwitnesses" to the city's
past and present sounds. They recorded background "keynote
sounds" like the steady pulse of traffic, "signal sounds" like train
whistles and other alerts, and finally the "soundmarks" unique
to Vancouver, such as the pealing bells of Holy Rosary Cathedral
and the aluminum "heritage horns" tooting the first four notes of
"O Canada" from atop the headquarters of BC Hydro.

In his 1977 book *The Tuning of the World*, Schafer worried deeply about how noise pollution might swamp the world's
unique soundscapes. But he deemed noise control "a negative
approach" to the problem. He hoped people would lend more
consideration to *all* the sounds around them, both wanted and
unwanted. "We must seek a way to make environmental acoustics a *positive* study program," he argued. "Which sounds do we
want to preserve, encourage, multiply?"

For decades thereafter, mainly academics, musicians, and
sound artists discussed, studied, and debated soundscapes. But
by the turn of the twenty-first century, these ideas had begun
creeping into the vocabularies of architects, urban planners,
ecologists, public health officials, and others who hoped to move
beyond the Whac-A-Mole approach to decibels and instead
think about sound as a potential resource. If we could consider
soundscapes more deeply and proactively, then maybe we could
harness them to our larger ambitions for more productive workplaces, healthier buildings, and cities with a mix of restorative
places, lively buzz, and easy connection.

The practical and philosophical challenges of such a shift are enormous. Despite the decibel's limitations, it provides a reliable, objective measure of acoustics that's woven into almost every regulation and building code pertaining to sound. Soundscapes, by contrast, celebrate subjectivity and unique contexts. Prioritizing them will require new tools, methods, and metrics, so we can listen to how different design choices will change the sonic experience within a planned but not-yet-constructed building and even predict how future soundscapes are likely to be perceived.

Ultimately, we need to foster a new mindset premised on the belief that our sound environments should be judged by much more than decibel counts, that quiet is only one worthy sonic goal among many, and that sound is a potential ally waiting to be mobilized. As Schafer put it, "The final question will be: is the soundscape of the world an indeterminate composition over which we have no control, or are *we* its composers and performers, responsible for giving it form and beauty?"

Cultivating better soundscapes was an audacious notion half a century ago, and it remains one today. But there's growing momentum toward making the attempt. The first step is to embrace the possibilities implicit in sounds and our prerogative as listeners to demand more from them.

The Long, Winding Road to a Less Noisy Future

The noise trap can feel immutable and inescapable. As technologies advance, new sounds supplant the old and become the next target of anti-noise movements. As a result, the fight against noise can sometimes seem more like a crusade against change and modernity rather than a campaign to reduce sounds that are doing actual harm.

Within this larger continuity, however, a critical shift is underway in the composition of daily noise. Specifically, a growing portion of the sounds hitting our ears are not the unavoidable sonic side effects of equipment as it functions or vehicles as they move. Instead, these sounds are intentional design features. Grafted onto everything from cars to phones to household appliances, these beeps, ringtones, and whooshes may collectively seem like the inescapable acoustic ambience of our time. But none of them are inevitable. All of these sounds were created by somebody, and like any design choice, they can be thoughtful or expedient, creative or lazy. Most important, they can be changed.

What constitutes noise in the future will remain as subjective as ever, but an expanding universe of designed sounds offers a huge opportunity for shaping soundscapes, if we can seize it by prioritizing the needs of listeners. Among the best examples is the chance to overhaul the sounds of city traffic. For nearly a century, vehicle noise has been one of the top sonic complaints. We could do things to absorb, block, and otherwise dampen the decibels of internal combustion, but the sounds themselves could not be extinguished, as they were part and parcel with mobility, commerce, and economic development.

Yet gas-powered transport is slowly but surely giving way to electric vehicles (EVs), which are so quiet* that sounds must

---

* Engine noises dominate the traffic soundscape at the lower speeds typical of city driving. Once cars accelerate above 30 miles per hour, the sounds of tires slapping pavement takes over, along with the whoosh of air rushing over the vehicle frame at high speed.

be *added* to them, so that pedestrians can hear them as they approach. According to the International Energy Agency, people bought nearly 14 million electric cars in 2023, accounting for about 18 percent of all cars sold, more than tripling the tally from 2020. Estimates for the future vary widely, but some project that the majority of all cars sold by 2030 will be EVs. As the massive global noise of internal combustion inexorably recedes, what comes next is up to us.

Beeps, Bleeps, and Bubbles

In December 2018, executives from the transport authority of London filed into a conference room at Anderson Acoustics. They were there to discuss the city's future soundscape, but they didn't realize that yet. The transport authority had recently made several bold pledges to upgrade its bus system—expanded and more accessible service, zero deaths due to collisions (down from about a dozen most years), and phased-in electric buses to help the system get to net-zero greenhouse gas emissions. The goal was to implement these changes by 2030.

The new electric buses would need an acoustic vehicle-alerting system (AVAS) to satisfy regulations and help the authority meet its safety goals. The original request for proposals stuck to the letter of the acoustic law regarding a "beacon sound" to alert the public to the presence of a moving bus. Namely, at speeds below 20 kilometers per hour (about 12.5 miles per hour), the warning needed to consist of two tones in specified ranges of pitch and decibels. One of these tones would rise or fall in frequency to indicate that a bus was speeding up or slowing down.

The folks at Anderson, however, led by senior consultant

Grant Waters, sensed a potential missed opportunity. Every day, millions of people ride London's iconic red buses. Street by street, these vehicles form an important thread in the fabric of the city. The alert sounds the transport authority chose would be amplified as thousands of buses regularly crisscrossed London, fundamentally changing the city's soundscape, especially as the growl of internal combustion faded.

"If you simply did what the regulation requires, the result would be very plain, boring, and ultimately unpleasant," Waters told me. Nevertheless, the transport authority was thinking about emissions and safety, not about city soundscapes. "When we first asked the transport authority what they actually wanted the buses to sound like," Waters recalled, "they said, 'Well, we don't really know.'"

To be fair, neither did the acousticians. They'd been trained to isolate and minimize environmental sounds such as those made by traffic, not to imagine new sound possibilities. They decided that the best way to brainstorm ideas was to stop thinking about bus sounds in isolation and consider them as part of the overall city soundscape. So, as people gathered in the conference room at Anderson Acoustics, they saw a big screen displaying a spring scene in a residential street of a busy area in London called Elephant and Castle where several major roads, railways, and subway lines converge. In the display, pedestrians wearing light jackets hurried past apartment blocks and newly leafed-out trees bathed in morning sun. In the foreground was a bus stop.

Speakers surrounding the assembled group played recordings of the neighborhood's actual urban hum, slightly subdued to approximate a near-term future with fewer internal combustion engines. Birds chirped as they flitted overhead, the footfalls of people on their way to school and work padded along, and a bicycle briefly whirred through. Into this sonic scene the Anderson team added electric buses arriving at the bus stop, one after another. The different alert sounds they made had been sampled from the current offerings of electric car manufacturers.

This first meeting was merely a conversation starter. No new sounds were crafted, and no contracts were awarded. The point was to start thinking beyond the regulations. While the primary job of the new bus sounds would be keeping people safe, they could do much more, as part of Londoners' daily lives and the city's acoustic identity. They could even reflect London's aspirations for the future. With all that in mind, what should the new buses sound like?

Answering that deceptively simple question would take years and a painstaking design process grounded in a lot more listening. There was initial enthusiasm for riffing on sounds that reflected London's heritage, such as Big Ben's chimes or the iconic refrain reminding Underground riders to "Mind the gap" between the train and the platform. These ideas soon gave way to others meant to invoke broader concepts, such as efficiency and sustainability.

There were some stumbles along the way. Early in the summer of 2019, the project team* invited Londoners to an initial stakeholders' workshop, where participants listened to a range of possible future bus sounds—some inspired by bubbles and bicycle spokes as well as a range of various bleeps. It did not go well. This workshop was meant to be a first-pass brainstorming ses-

---

* The sprawling AVAS project team included the international infrastructure consultancy AECOM and sound designers at Zelig Sound. Anderson did not officially join the team until later in 2019.

sion and a chance to think outside the regulatory box, the aural equivalent of a decorator showing clients fabric swatches and paint chips. Yet the invited listeners were given very little context for judging the sounds, and their reactions ranged from perplexed to disdainful—"very spaceshippy," "irritating," "ghastly."

"A lot of people were saying, 'These sounds aren't relevant,'" Waters recalled. "'We don't want our buses to sound like bubbles!'"

In retrospect, the causes of this debacle seemed clear to him. You couldn't just present people with sounds and ask them which ones they liked best, nor whether bubbles or bleeps sounded more like a bus, because a lifetime of listening has taught people that buses sound like neither of those things. The project team needed to break through this inertia step by step, by encouraging participation in the entire project and getting people to embrace the possibilities of new sounds and the importance of their role in crafting them. To that end, subsequent workshops didn't start with sounds but with an explanation of the mission, including the benefits of an electric bus fleet, the public safety role of the new alerts, and an invitation not only to give their opinions on potential new bus sounds but to codesign them and to help shape the future soundscape of their city.

"We needed to convey the whole story to explain why we were doing this and bring people along on the journey," said Waters.

The workshops included people who would need the buses' alert sounds the most, such as the visually impaired, cyclists, and the elderly. People with other concerns, such as the bus drivers who would hear the new sounds constantly, day in and day out, were also brought on board. Together they entered simulated sonic worlds: quiet residential roads, busy streets, and bus depots. The workshop attendees listened to phantom buses making a wide range of sounds while coming, going, and idling in these various scenes.

As the work progressed, the sound designers on the project tapped into research by experts in psychoacoustics—who study the links between a sound's acoustic properties, such as changes in frequency, cadence, and decibel levels, and how the sound is perceived. For example, psychoacoustic studies have found that listeners will judge a sound to be "sharper" when it has more concentrated energy in the higher frequencies, while rapid fluctuations make sounds "rougher." A whistle, squeal, or hum, with a dominant narrow-frequency sound, seems more "tonal." Other changes to a sound's mix of frequencies or the suddenness of its audible beginning (attack) and end (decay) can make it seem "rounder," "warmer," "brighter," and so forth.

Workshop participants weren't simply presented with designed sounds and asked to pick favorites. They were instead asked perceptual questions, such as which sounds provided a better feeling of safety and which ones suggested that something as big as a bus was approaching. They were asked how candidate sounds could be improved and made more useful, and how the most useful sounds could be made more pleasant. Based on feedback and suggestions, the sound designers would make real-time tweaks before asking participants to listen again and compare notes.

The influence these stakeholders had on the sounds that were eventually chosen was substantial. It was the workshop participants, for instance, who insisted on an idling sound for a stationary bus, even though regulations required only a beacon sound for a bus in motion. They also took part in subsequent road tests of top-candidate sounds on a single bus route. The new alerts would automatically become quieter at night or when the bus traveled through residential streets (tapping into an existing geolocation system that capped bus speeds at a road's legal limit). And the field trials led to further refinements, such as adjusting the position of a speaker to lower the sound volume inside the cabin and using only forward-facing speakers to avoid confusion when several buses queued up at a bus stop or terminal.

At the end of this protracted process, the winning sounds were a nearly unanimous choice—a plunking beacon of electronic droplets that rose and fell in pitch as the bus changed speeds and faded out entirely beyond 20 kilometers per hour, along with an idling sound of subtly musical, gently modulated chords reminiscent of a diesel engine's churn. Equipped with the new sounds, London's fleet of electric buses has since expanded rapidly, reaching nearly a thousand buses in early 2023.

Yet despite the care and wide-ranging input that went into crafting these bus alerts, there's nothing sacrosanct about them. Some people won't like the new sounds from the jump. For others, the alerts will inevitably wear out their welcome over time, especially as the larger soundscape of London evolves. As the increase in electric vehicles leads to quieter streets, bus sounds could be made quieter or changed entirely. A key piece of creating sounds for the future is recognizing their impermanence and making them amenable to constant evolution.

Avoiding Sonic Trash

Granted, many of the designed sounds in our lives aren't replacing legacy noises. Instead, they are jostling their way into an already crowded soundscape and vying for another slice of our overstretched attention. For these sounds, the potential to devolve into noise is especially high, according to Joel Beckerman, the founder of Made Music, a sonic branding studio in New York City. As he put it, "People can quickly start to hate a sound when it doesn't help them."

Indeed, Beckerman calls "sound for the sake of sound" a kind of "sonic trash"—an offense worse than plain old noise, which is the unfortunate by-product of something useful, such as a subway or a jackhammer. He thinks too many brand managers treat increasingly noisy soundscapes as competition for customer attention and consider any sound-free moments to be missed opportunities. Hence, the endless jingles and the relentless fanfare of smartphone apps. This is why, despite being in the sound-making business, Beckerman extols the value of silence. In any composition, silences are much more than empty space between notes, he pointed out; they mark time, build anticipation, and create resonance.

"All we have to work with is sound and silence," Beckerman said. "If everything has sound, then nothing has meaning."

Made Music's core business is composing sonic logos and music that brands weave into advertisements, products, and sponsored experiences. According to Beckerman, an important step in their design process is pruning sounds from a work in progress to see if a client notices the omission. "If they don't miss a sound," he said, "then it never should have been there."

In Beckerman's estimation, the opposite of sonic trash are the helpful sounds designed specifically to *ease* our cognitive load by orienting and guiding us or by letting us know that we've accomplished a task, such as completing a digital payment or submitting an online form. He played me a few digital-interface sounds under development for an unnamed client—one evoked a flywheel for scrolling through options, a "zip" page flip for opening an application, and a plunking droplet for depositing a file.

Out of context, these sonic tidbits were hard to judge and up for the usual subjective debate. But as Beckerman introduced the sounds, he talked about design choices made to boost a sound's utility for listeners. He noted how meaning and emotion can be carried by different timbres, rhythms, and other little hints, such as acceleration implied by a rising pitch or an email dispatched with a satisfying whoosh.

A sound's usefulness is not based solely on the importance of its message but also on its ability to give listeners the right amount of information at the right time, with the least amount of cognitive friction. Since 2018, Made Music has worked to strike that balance by partnering with a New Hampshire startup called Sentient Decision Science, which seeks to uncover our deep emotional reactions to sounds by using the implicit association test (IAT). Developed in the 1990s by psychologists at the University of Washington, the University of Virginia, and Harvard University, the IAT can purportedly spot people's subconscious biases by comparing their response times when pairing rapid-fire stimuli (sounds, in this case) with value judgments such as good versus bad, welcoming versus aversive, and so on. For Sentient and Made Music, the rationale for using the test is that some of our strongest reactions to sounds are automatic and below our conscious awareness; they would therefore be missed by asking people direct questions about what they're hearing. For example, we might consciously assume that a sound needs to be jarring and unpleasant to convey urgency, but a 2018 Sentient study found otherwise.

They used the IAT with about nine hundred people to assess the emotional appeal of ten "naturally occurring" sounds such as birdsong, a scream of pain, rain, and a baby's laugh, along with ten designed sounds emitted by home security systems, microwaves, weather apps, and so on. They found that a sound's emotional appeal did not need to match the tenor of its message. Consider two severe-weather alerts that Sentient tested—one from the Emergency Broadcast System scored second worst on the scale of emotional appeal, right behind a scream of pain and edging out nails on a chalkboard in the ranks of unappealing sounds. By contrast, a severe-weather notification from the Weather Channel was ranked among the most pleasant sounds, behind only applause, birdsong, and a baby's laugh, and beating out wind chimes by some distance.

These findings should be taken with a grain of salt. Sentient is, of course, in the business of selling its insights, and the full scope of its methods and analysis are proprietary and often kept from public and peer review. Still, simply asking these questions about sound is a victory for our ears. No sound, no matter how expertly researched and crafted, will please everyone, but the risk that it will be perceived as noise skyrockets when a new sound is foisted on our ears without forethought.

Whenever we introduce new sounds, Beckerman noted, "we're either improving a person's experience or making it worse. There is no neutral."

Despite their rapid spread, designed sounds remain only a sliver of the noise in our lives. They are, however, a worthwhile place to start distinguishing sound and noise based on how they meet listeners' needs rather than where they register on a decibel chart. If our focus can expand beyond the shushing of loudness, then we can devote more thought and energy into minimizing sonic trash and striving for soundscapes that make our lives easier rather than harder.

7

ALL THE MACHINES THAT ALL GO BEEP

SOLVING SIGNAL OVERLOAD

Introducing her online exploration of New York City noises from a bygone era, titled "The Roaring Twenties," Princeton University historian Emily Thompson contrasted the bustling brick-and-mortar din of jazz-age New York—the sounds of steam shovels, factory sirens, and elevated trains—with today's "online cacophony" emanating from text message threads, email inboxes, and social media scrolls.

After observing that the popular uproar over audible noise has ebbed and flowed over the past century, Thompson noted the recent resurgence in complaints about audible noise and media coverage of the issue and asked, "Why do we now once again find noise so compelling?"

Her answer: "a curious kind of nostalgia." While acknowledging that noise can cause "very real distress," Thompson suggested that renewed attention to the problem may be spurred by "a complicated longing for the old-fashioned problem of physical sounds." Put another way, people who feeling overwhelmed by the crush of newfangled digital "noise" may find it simpler to vent frustration on the more familiar problem of audible noise emanating from the analog world, whether it's traffic, construction, or chatty office mates.

But perhaps our rising concerns about noise are not so much nostalgic as cumulative. After all, audible noise and the "online cacophony" are more than metaphorically intertwined. For one thing, many new sounds are the audible beacons of our sprawling digital networks and interconnected gadgets. For instance, every new text, social media engagement, or app notification is signaled with sounds (unless we explicitly silence them). More generally, if we swap "unwanted signals" for "unwanted sounds," then we cover both types of noise and find that the same cognitive filtering is required to fend off the dross—audible or otherwise.

By 2019, a *Boston Globe* writer complained that we might have reached "peak beep," thanks to "the proliferation of nervous Nellie machines determined to update us on every development," no matter how trivial, with a growing chorus of beeps, pings, and chimes.

Whether or not peak beep is upon us, there's no denying that we're increasingly surrounded by gadgets and gizmos that, like toddlers tugging at our sleeve, constantly have something they *need* to tell us right now. It's not just our washing machines, dishwashers, and coffeemakers chiming in about their completed tasks or the incessant phone and smartwatch pings from app updates, breaking news, and social media. Now "smart homes" alert us about everything from thermostats and light dimmers to the contents of the refrigerator, suggesting that maybe it's time to buy more milk. As a result, one of the biggest noise challenges of our time concerns cognitive overload. For both designing new sounds and rethinking entire soundscapes, we must keep our overburdened attention span in mind. The late computer scientist Mark Weiser predicted this problem of attentional overload in the late 1980s. Weiser worked at the Palo Alto Research Center (PARC), an exploratory outpost of the Xerox Corporation, where innovators in computer science lounged in beanbag chairs as they debated big visionary ideas, which they and their colleagues scribbled on giant whiteboards. At a time when people were still adjusting to the advent of personal computers, Weiser foresaw a coming age of "ubiquitous computing," in which people interacted with many small interconnected devices that he called, "pads, tabs, and boards."

By 1996, when email and the internet had gone mainstream but well before smartphones were our constant companions, Weiser and PARC's then director, John Seely Brown, wrote a landmark paper calling for "calm technology."

"If computers are everywhere they better stay out of the way," they wrote. These devices should not constantly shout, "Pay attention to me!"

A calm technology would insert itself into our focus only when necessary to help or guide us. Gentle nudges and timely reminders would help us feel less stressed and more in control rather than harried and frazzled by constant pestering. The problem of "unwanted signals" isn't simply annoying or frustrating; it's also mentally draining. Our minds yearn to declutter. We wearily skim and delete our way through spam-filled inboxes and silence our phones, despite the risk of missing important messages from a colleague or spouse. In much the same way, the more audible signals clamor for a sliver of our overstretched attention, the stingier we become with it, triggering an auditory arms race that further erodes our willingness to listen.

When Alarms Cry Wolf

There are few places where the hazards of this vicious cycle of noise are more evident—and more consequential—than in hospitals, where the number of alarms calling out to clinicians has skyrocketed in recent decades. From drug-infusion pumps to heart-rate monitors to low-battery warnings to bed alarms firing off at routine body shifts, many hospitals have tallied hundreds of alarms per patient per day. Most of these alerts are either false or inconsequential.

This auditory onslaught distresses patients and disrupts their sleep, and its effects on the people caring for them can be even more perilous. Clinicians' attentional resources are not unlimited, and neither is their capacity to remain on constant high alert. A flood of false and incidental alarms hastens burnout and triggers a cry-wolf mindset, which can slow responses to real emergencies while increasing the likelihood of medical mistakes.

A 2015 study at the Children's Hospital of Philadelphia demonstrated the risks of "alarm fatigue" in both the general ward and the pediatric intensive-care unit. Researchers obtained permissions to set up GoPro cameras at nursing stations and in patients' rooms for long-term monitoring. After reviewing thousands of alarms across hundreds of hours of video, they concluded that only 13 percent of intensive-care alarms and a scant 1 percent of general ward alarms warranted clinical intervention or consultation. By contrast, the vast majority were "nuisance alarms" that were either clinically irrelevant or simply false, such as an alert sounding when a baby dislodged a monitor by waving its tiny arms and legs. When the researchers tracked clinicians' response times to alarms requiring medical intervention, they found that the greater the number of nuisance alarms over the two hours prior, the slower clinicians reacted when it mattered.

No kids were hurt due to the delayed responses observed in this study, but analyses of nationwide hospital data suggest that alarm overload is a real threat to patients' safety. A four-month audit by the Food and Drug Administration in 2010 uncovered seventy-three alarm-related deaths nationwide, and a ten-year review in 2014 blamed 844 injuries on errors related to alarms. In 2013, the American Association of Critical Care Nurses (AACN) declared alarm fatigue a danger to patients.

Hospitals have mainly tried to solve the problem of alert overload with smarter management of alarms—by rigorously evaluating whether a patient needs to be hooked to every monitor while carefully tweaking the threshold for when nurses and doctors must be alerted. Before making such changes, they conduct pilot studies to ensure that patient safety isn't compromised. The Children's Hospital of Philadelphia, for example, made alarm fatigue a focus of their Patient Safety Learning Lab. In one intervention, they lengthened the automatic delays in blood-oxygen alarms, which are meant to account for the frequent episodes when blood-oxygen levels dip below normal before quickly rebounding. By extending those delays from 10 to 30 seconds, the hospital cut nuisance alarms by more than 80 percent, according to a 2018 pilot study. They made these moves cautiously, starting with a single ward and monitoring it for any upticks in code blue emergencies before widening the implementation.

While better alarm management makes a difference, it does not solve the whole problem. In 2023, the AACN published a follow-up article in its journal, *Advanced Critical Care*, titled "Ten Years Later, Alarm Fatigue Is Still a Safety Concern." The authors applauded the progress that had been made in acknowledging the problem of proliferating beeps, and they highlighted the successes of specific interventions. But ultimately, they concluded that the battle was far from won.

"Alarm fatigue continues to be an issue," they wrote, "[contributing] to missed alarms and medical errors that result in patient death, increased clinical workload and burnout, and interference with patient recovery."

As they confront alarm fatigue, hospitals face a soundscape problem that needs a holistic soundscape fix. Alarms are essential to clinicians in their work and critical to patient safety. As such, the goal should not simply be to minimize the number of alarms but to optimize them, in the spirit of calm technology—so that alerts can cut through noise rather than add to it. Ideas that reflect the goals of calm technology include patient monitors that can sense and react to the prevailing soundscape and more helpful sounds that demand less mental bandwidth.

Smarter Sounds

There's a lot riding on all those hospital beeps. They must be learned easily, noticed immediately, and convey their message clearly. Above all else, they must not fail. An alarm that gets lost in the background din adds to auditory clutter and endangers patients. Therefore, the conventional wisdom has been that louder alarms are always better, or at least safer, according to Joe Schlesinger, the Vanderbilt anesthesiologist who collaborated on research into the CanaryBox for noisy operating rooms.

"The thinking is, well, no one is dropping dead from our alarms. If we make sure they're louder than the ambient sound levels, then we know people are not going to miss them," he said.

"But, I thought, why? Why do alarms have to be so loud and annoying?"

Schlesinger is a self-proclaimed music geek who majored in jazz piano performance at Loyola University and put himself through medical school by playing gigs in New Orleans clubs and performing with the showband aboard Carnival Cruise Line ships. Music informs both his clinical practice and his passion for fighting hospital noise.*

"There are so many ways sounds can seize your attention and brim with meaning without being the loudest thing out there," he said. Schlesinger stressed the complexity of sounds, and the myriad combinations of timbre, harmony, and rhythm. To illustrate, he cued up some Afro-Cuban music on his computer and described the quickening *clave* rhythms which he hears echoes of in irregular heartbeats. At the very least, he wondered why medical devices couldn't exercise a little discretion by sensing changes in a hospital soundscape and adjusting the loudness of their alarms accordingly?

Before he could answer that question, Schlesinger had to determine just how loud alarms really needed to be, relative to background noise, to remain effective. To find out, he recruited about thirty anesthesiologists to take a turn in Vanderbilt's three-story-high anechoic chamber—a thickly insulated room designed to eliminate echoes—and confront a series of simulated medical emergencies. The chamber's walls and cathedral ceiling are covered with foam wedges that jut out like enormous beige teeth and extend about 10 feet below a mesh-metal floor into a gaping, sound-swallowing maw. At the chamber's center, a

---

* Schlesinger has a secondary faculty appointment with Vanderbilt's Blair School of Music.

black-leather pedestal chair is ringed, at head height, by dozens of small speakers suspended from above.

In Schlesinger's study, doctors sat in the pedestal chair as the surrounding speakers pumped out ambient hospital sounds at a steady 60 decibels. Their main task was to track a suite of audible vital-signs monitors, which would randomly indicate different emergencies requiring specific drug interventions. The doctor could select the drugs via a keypad on the chair's arm. To heighten the challenge, subjects faced a couple of distractor tasks—spotting a flicker of yellow light in the chamber and listening for a target phrase in a jumble of three voices. Over hundreds of trials, the loudness of the alarms ranged widely, from 4 decibels louder to 30 decibels quieter than the background sounds. At what point would the ambient noise swamp the alarm signal and impair the doctors' emergency decision-making? Surprisingly, there was no drop-off in the speed and accuracy of the doctors' drug choices until the alarms became more than 11 decibels *quieter* than the background sounds.

These results, published in 2018, led Schlesinger's team to devise and patent a gizmo called DASH (Dynamic Alarm Systems for Hospitals) that could continuously sample the level of ambient sound and modulate the loudness of alarms to maintain a desired ratio of signal to noise. But up to this point, device makers have shown little appetite for adding an automated volume knob to their alarms. So, for now, the framed patent simply adorns Schlesinger's office wall.

Schlesinger's lab often takes an engineering approach to subduing sonic foes. Another of the projects they developed was an in-ear device for intensive-care patients that filters out alarm-specific frequencies to minimize the distress caused by constant alerts and help patients sleep, while at the same time keeping intact the sound patterns that are typical of speech, thus enabling easy communication with caregivers and loved ones. Also, in partnership with Jeremy Cooperstock, an expert in human-computer interaction and professor of electrical and computer engineering at McGill University in Montreal, Schlesinger has been experimenting with haptic alarms. Inspired by the vibrational alerts of pagers and mobile phones, they aim to reduce clinicians' distraction during surgeries by replacing some auditory alarms with a silent, gentle touch.

Fighting Sonic Inertia

The battle against alarm fatigue is not simply about volume control but also about revamping hospital soundscapes to better support clinicians and patients. Momentum for this larger cause continues to build on multiple fronts.

Among those working to extricate hospitals from the noisy trap of their own making is Yoko Sen, a sound designer and composer of breathy ambient electronic music in New York City. About a decade ago, she cofounded Sen Sound with her husband, after an extended hospital stay. Through countless interviews and workshops with patients, clinicians, and medical device makers, Sen and her team encourage people to think expansively beyond the beeps, not only focusing on the hospital noises they want to escape but considering how to create a better soundscape for healing.

During Sen's hospitalization, she lay surrounded by machines beeping with unflagging urgency, and she thought about a neuroscience study she'd read, which suggested that hearing is the last sense to go before we die. Would this piercing chorus of beeps be her final sendoff? It seemed so tragic. At one point, an alert on her bedside monitor sounded so shrill and persistent that her husband flagged down a nurse for help.

"Oh, don't worry," the nurse reassured him. "That thing just beeps." Telling this story, Sen laughs at the notion of a machine that "just beeps," without meaning or apparent purpose beyond sounding its unceasing electronic yawp. But the vignette also captures the neglect of the wider soundscape that lies at the heart of alarm overload. Its cacophony is a result of the decades during which hospitals accumulated sound-emitting technologies with little regard to the mounting cognitive burden they might place on staff and patients. In 2018, for example, a lead product developer at Philips, a prominent medical device company, contacted Sen after watching a talk she gave and invited her to work with his team. He admitted that his company had largely ignored the alarm sounds their patient monitors made, despite devoting decades of engineering and countless dollars to research and development of the machines' other aspects. Eventually, he traced the current alarm sounds to a cassette tape dating from 1981.

This is an example of "sonic inertia," and it wasn't surprising to Sen. Constant beeps might keep patients awake and anxious, and they might stress out busy nurses and doctors, but as long as they successfully snagged clinicians' attention, they worked, and the sounds they made were left alone. Sen's first challenge, therefore, was to raise expectations of alarms beyond basic functionality. Collaborating closely with the Philips product-design team, she played the company's then-current alarm sounds to gatherings of doctors and nurses and asked them, "If the patient monitor was a person, who would it be to you, based on these sounds?"

"A drill sergeant," somebody said. Others suggested a dictator, an ignored boss, or a petulant toddler. When asked what they might prefer, the clinicians said they wanted the monitor to sound more like a coach, a friend, or some other helpful and supportive person.

"I wish it to be a colleague," one doctor ventured, "with the same interests as myself: to do what's best for the patient."

The next step was a series of virtual workshops to solicit broader thoughts on hospital alarms from both clinicians and laypeople in a dozen countries. The project team dissected the transcripts of these sessions, clustered common themes, and ultimately distilled eight criteria for a successful alarm. Half of them related to "functionality," such as being simple to learn and locate in space, easy to distinguish amid background noise, and quick to stimulate a response. The other half were measures related to "sensibility," including how startling, aggravating, fatiguing, or distracting the current sounds were. With these criteria as a guide, Sen Sound and Philips's product designers started reworking the sounds for low-, medium-, and high-priority alarms.

Clinicians had said they wanted the low- and medium-priority alarms, which they heard most often, to be less aggressive, so the team slowed the pacing of the beeps and made them softer and "rounder," more like chimes with lingering notes. Medium-priority alarms were distinguished from low-priority ones by making them a bit more percussive and higher pitched, "like a gentle tap on the shoulder." High-priority alarms kept the original pacing of beeps, but the timbre was tweaked to make it less shrill and harsh—clinicians said they wanted to be "warned without being jolted."

The project team then gathered feedback via two rounds of online surveys embedded with sounds—first they tested a handful of alternative alarms against each other and then pitted the winning alternatives for low-, medium-, and high-priority alerts against the originals. They asked listeners to give their preferences and to rate the alarms on the eight criteria.

This led to a surprising discovery. Everybody had expected significant tradeoffs between functionality and sensibility. They assumed that the acoustic qualities that grab attention and prompt immediate action would be opposite in nature to those that make sounds softer, gentler, warmer, or smoother. Yet survey respondents mostly ranked their preferred sounds higher in *both* functionality and sensibility. And they preferred the new sounds to the originals overall. Philips adopted the redesigned sounds in 2023.

Who's Sorry Now? A Soundscape Saga

Meanwhile, a separate international collaboration of researchers and clinicians was plotting even more radical changes to the hospital soundscape by introducing totally new alarm sounds that completely bypassed the beeps. Their story, which follows, is a tale of ruin and redemption, creative destruction, and the struggle to coax meaning from cacophony.

It all began with a joke gone wrong. At an after-dinner presentation during the 1990 annual meeting of the American Society of Anesthesiologists, Frank Block Jr. announced that he'd devised new sounds for six alarms—ventilation, drug delivery, temperature, oxygen, cardiovascular, and artificial perfusion (blood pumps)—known in the critical care world as "the six ways people die."

Born in Atlanta and now retired, Block is a raconteur with a rollicking Georgia accent who earned an undergraduate degree in music on his way to medical school. The new alarms he proposed were a selection of well-known songs such as "I Left My Heart in San Francisco" for the cardiovascular alert, the "Chariots of Fire" theme for temperature, and "Blowing in the Wind" for ventilation. These alarms would be a cinch to learn and instantly identifiable in an emergency, Block insisted in his presentation. He passed out quiz sheets to his audience, then he played the songs again in a random order, asking listeners to match each tune with its intended message.

"Afterward, people kept running up to me and saying, 'Are you kidding? Is this a joke? Is this for real?'" Block recalled. "Quite honestly, I didn't know."

In truth, it was both, plus a bit of sabotage. For much of the twentieth century, as patient monitors proliferated, few guidelines governed the sounds they made until 1986, when the International Organization for Standardization (ISO) finally decided to tame the auditory chaos by commissioning the first standardized alarm sounds. They hired a psychoacoustics expert at Cambridge University (UK) named Roy Patterson, who had previously designed alarms for the British military and commercial airplane cockpits. For the hospital alarm project, Patterson enlisted the help of a postdoctoral researcher named Judy Edworthy, who is now a professor of applied psychology at the University of Plymouth (UK).

To cover the six ways people die, plus a general alarm, Patterson and Edworthy composed seven synthesized melodies of three to nine notes each, with two possible tempos (slower for a cautionary alarm and faster to indicate an emergency). The leading journals and societies for anesthesiologists then circulated a toll-free phone number that people could call to hear the proposed alarms. "Everybody dialed that number and listened to these alarms," recounted Block, who had worked on medical monitoring technologies during a fellowship at Duke University. "We all said, 'These are ridiculous! These are terrible! We've got to do something!'"

Hence, the pop-song counterattack. Block was serious enough about it to publish a follow-up paper noting that more than half of the anesthesiologists in his after-dinner audience had correctly matched all six song alarms with the proper emergency event after only one hearing. The paper even suggested workarounds for copyright issues and varied instrumentation to help clinicians pinpoint the precise location of a particular alarm in intensive-care units and other locations where several patients were tethered to multiple monitors.

By the late 1990s, Patterson and Edworthy's alarms were dead in the water, thanks in part to the widespread derision Block fomented. But the ISO didn't opt for Block's alarm playlist either. In a compromise that pleased no one, the standard keepers instructed a joint working group on alarms to keep Patterson and Edworthy's tempo cues for urgency, but little else. Melodies were fine, but they needed to be much simpler—only three notes for cautionary alarms and five notes for emergencies. Finally, in a coup de grâce, the ISO asked Block to help compose the stunted tunes.

"To come up with six or eight different alarm sounds and make them distinguishable in three notes is pretty much impossible," Block told me. Still, the group did their best to freight each composition with as much meaning as three notes could carry. For ventilation, they borrowed from the NBC chime, with tones that rose and fell like the action of lungs. The oxygen warning made a slow descent because only falling levels of oxygen worry clinicians. The notes of the cardiovascular alert, by contrast, rose like a call to arms.

Amazingly, nobody tested these truncated tunes for how easy they were to learn and pinpoint in a busy hospital before they appeared as ISO-approved options for device makers in 2003. When several researchers eventually *did* test the new alarms, Block recalled, "No big surprise, they were terrible!"

Deepening Block's regret, he had subsequently met Patterson, who had explained the mnemonics at the core of the alarms Block had sabotaged. For instance, a six-note staccato, beep-beep-beep-bup-bup-bup, was a tonal rendition of "car-di-o-vas-cu-lar." The ventilation alarm was a slower five-note pattern of high-low-high, a brief rest, and then a two-note descent for "Ven-til-ate Pa-tient."

Block subsequently apologized publicly for derailing Patterson and Edworthy's efforts. And while Patterson has long since moved on from alarm sounds, Edworthy stayed in the alarm game, thanks in part to Block's asking her to try again in 2014, when he cochaired an alarms committee created by the Association for the Advancement of Medical Instrumentation—an American nonprofit that creates standards for health-care technology. Knowing how fickle the collective ear could be, Edworthy said she would consider the project only if there was ample funding to thoroughly test her alarms and prove their superiority *before* they faced the judgment of the wider medical community. It took a couple of years, but by 2016, Block's committee had raised about $100,000 of funding for Edworthy's alarm research, and so she set to work.

She knew from previous research that two key ingredients would make acoustic alerts easier to learn. The first was concreteness: a sound with strong real-world links will trigger automatic associations, such as honking with rush hour, gulls with the seashore, a pipe organ with church. The second ingredient was uniqueness: while sets of alarms should share an acoustic kinship, they also needed to sonically stand apart. Beeps fell short in both dimensions; they were both abstract and difficult to distinguish.

With the help of collaborators in Europe and the United States, Edworthy came up with "auditory icons"—sounds with strong metaphorical connections to the six ways people die. The ventilation alarm, for instance, sounds like a bellows inflating and deflating, while the cardiovascular alarm resembles a digitized heartbeat. The low-drug warning, meanwhile, sounds like the rattle of a nearly empty bottle of pills. They added an alarm for power failure and dropped the general alarm in favor of an attention-grabbing "pointer" of three or five pulses (for medium and high priority) that would be tacked onto each auditory icon, with a tempo to match the alarm's urgency. They then spent two years testing the auditory icons against existing alarm sounds, including those created by Block and company.

In both lab experiments and simulated medical emergencies, the auditory icons bested the beeps. In one such study, for instance, standard alarms were correctly identified only 43 percent of the time, compared to 88 percent for the icons. The icons also prompted the proper medical response 3 seconds faster and caused less frustration and fatigue among clinicians in a simulated ICU. Nevertheless, Edworthy steeled herself for pushback from the ISO's alarm committee when they formally considered auditory icons for a 2020 revision of the alarm-sound standards.

"I thought, oh, they won't like these," she recalled. "They'll say, 'We can't possibly use the sound of a rattling pill box to signify drugs. It's too silly.'"

In the end, however, the weight of her evidence won the committee over, and auditory icons became an option in the new

standards published that summer. Edworthy and the committee agreed that the icons' performance benchmarks should be
included with the new standards to guide the composition of
future alarm sounds that might equal or better her creations.
She would welcome more sonic innovation as a sign that people
were taking hospital soundscapes seriously enough to overcome
decades of neglect, which had let noise grow like auditory mold.
After all, she said, "You can always design a better alarm."

A Final Sound

Solutions to hospital noise that have staying power will move
beyond squashing problem sounds and proactively consider the
bigger sonic picture: soundscapes matter for healing. Noise is
the top complaint in patient-satisfaction surveys created by the
Centers for Medicare and Medicaid Services, which rate hospitals on everything from pain management to cleanliness. Since
2012, a hospital's survey results have determined a sliver of its
Medicare reimbursement dollars, and that monetary incentive
has spurred more experiments with coordinated "quiet times"
by steering nonessential tests and consults to specified hours,
closing patients' doors, and putting up signage about the importance of peacefulness and rest for patients' recovery.

Studies find that quiet times make for a more tranquil hospital soundscape, and that improves patients' sleep. Knock-
on effects include reducing the use of sedatives and the risk of
delirium in critically ill patients, which can help shorten stays
in intensive care. The quiet times also tend to lower clinicians'
stress. But these restful interludes must be constantly defended
as staff turns over, workloads mount, departments reorganize,

and higher priorities intrude. There's a fragility to simply shushing business as usual.

In a 2016 presentation at Medicine X, a health-care innovation conference at Stanford University, Yoko Sen, the sound designer who would go on to help rework Philips' alarms, noted that change is not easy, and low expectations can become a survival strategy. She then shared the story of a palliative care nurse she'd met who sometimes played "meaningful music" in the rooms of terminally ill patients being taken off life support. Sen closed her presentation by inviting her audience to reach out later with their personal thoughts on the final sounds they would like to hear in this life. She recorded their responses and wove excerpts into an ethereal musical composition, along with some of the sounds themselves, which she performed at the conference's closing ceremony.

With the lights dimmed and the room immersed in a projection of rippling dark-blue and slate-gray, voices talked about the sounds of water as a source of life, a transformative force, or the reassurance of breaking waves reaching out from a vast, unfathomable ocean. Others spoke about hearing the voices and laughter of loved ones, a chorus of birds to kindle "that feeling of the morning when I'm waking up to something new," or simply a rhythmic beat strong enough to "carry me to whatever's next." For nearly five minutes, the darkened space echoed with final sonic wishes. Not one beep could be heard among them.

Sen's performance posed a musical challenge to sonic inertia and the false choice between noise and silence. Fatigued by signal overload and surrounded by machines that "just beep," we may find that our first instinct is to shut out as many sonic intrusions as possible and seek the temporary refuge of silence. But not all the noise cluttering our minds is of the audible variety.

Maybe a walk around the block or a gathering with friends—as noisy as these activities may be—will do a lot more for our well-being than sequestering ourselves and quietly scrolling through our social media feeds.

We should also feel empowered to demand more from sounds beyond those we wish to escape. We can push for calmer technology, with smarter acoustic options that lighten our cognitive load. We can also pose bigger questions about what we want from our soundscapes—be they for healing, learning, productivity, or play.

8

SOUND CASTLES

BETTER LIVING THROUGH LISTENING

From deafening restaurants to offices plagued by distraction, businesses spend billions of dollars annually on sound-absorbing panels, insulation, and other measures to counteract noise problems they could have avoided if they'd taken sound seriously from the start.

All this blundering into cacophony isn't simply a matter of shortsighted planning. It's symptomatic of a reactive mindset that understands noise primarily as loudness. This way of thinking doesn't entertain the possibility that soundscapes can be designed to enhance an environment beyond keeping noise below specified decibel thresholds.

Good indoor acoustics cannot, however, be defined by a single metric. They must be calibrated to a building's architecture, materials, purpose, and surroundings. Ideally, the resulting soundscape will not merely be free of major noise problems but will actually support the needs of the building's occupants. For example, we would likely favor different sonic backdrops for a celebratory night out, a religious service, or a busy day of focused work and tight deadlines. But too often, soundscapes are an afterthought in building design, except for checking the boxes of regulatory requirements and for specialized projects such as libraries and performance halls. In the visuals-first field of architecture, the drama of a vaulted ceiling or the appealing expanse of an open floorplan are easier to imagine than how these spaces will sound when filled with chatty coworkers, noisy machines, or the clatter and commotion of a restaurant kitchen.

Fortunately, both the means and the motivation for soundscape awareness are slowly infiltrating mainstream building design. Thanks to advances in aural simulation, both architects and their clients can "listen" to their projects before they're built and make tweaks to achieve soundscapes that better suit their purposes. Paired with these tools is a growing push to make buildings healthier places, and that includes less noise. We spend more than 90 percent of our lives indoors,* and we could make that time sound a lot better.

Architecture of the Ear

Few places are as noise-infamous as restaurants. While nobody enjoys eating in oppressive silence, the far more common problem for diners is the need to shout at one another and the waitstaff to be heard. A 2018 study of noise sampled at thousands of Manhattan restaurants found that the average measure was 77 decibels, with a quarter of the restaurants topping 81 deci-

---

*   Specifically, 93 percent for Americans, according to the EPA.

bels; only 10 percent dipped below 70 decibels (what the authors defined as "quiet"). That same year, noise claimed the top spot in complaints about restaurants, ranked by Zagat's annual Dining Trends survey. Noise beat out poor service, lousy food, and high prices. It wasn't always this way. By most accounts, the unbalancing of restaurant acoustics began in the late 1990s, when owners started favoring a modern industrial look—hard, sound-reflective surfaces, open kitchens, and exposed ceilings were in, and carpets, upholstery, drapes, and other stodgy sound absorbers were out. Eager for a high-energy ambience, restaurateurs also cranked up the music, and guests reflexively raised their voices to be heard above the din, triggering a vicious cycle of noise known as the Lombard effect, which compounded the cacophony. Writing about the resulting "Great Noise Boom" in *New York Magazine*, the restaurant critic Adam Platt described how incidental and deliberate loudness flowed together and reinforced each other until "front-of-the-house culture was slowly buried in a wall of sound."

Nevertheless, it's worth taking a closer look at one eatery that followed most of the noise-inducing trends and yet, somehow, managed to escape the same ear-bursting fate. The atrium café at the heart of Boston's Museum of Fine Arts is awash in natural light that pours through its towering glass walls and ceiling. The atrium in which the café operates was built in 2010 and is nestled between the original nineteenth-century brick facade and a three-story limestone wall bisected by a cantilevered staircase leading to the Art of the Americas wing, which was added the same year. Visually, the place is stunning. Acoustically, it *should be* a nightmare—a wide-open reverberant box crammed with people, including kids, just released from the hushed confines of galleries, plus hundreds of diners eating and chatting at tables that sprawl across the space, with no dividers in sight.

Instead, the restaurant is a harbinger of what soundscape design can accomplish, with some foresight and the right tools. On a recent Sunday afternoon, the jazz trumpet of Miles Davis flowed down from hidden speakers and mixed with the bustling crowd, the clinking cutlery, and the dishes being dropped into busing tubs. Yet all these sonic ingredients churned up just enough acoustic energy to support comfortable conversation, averaging 64 decibels when measured from one of the tables.

How did the café, with all its hard surfaces, large crowds, piped-in music, and lack of interior walls, pull off this acoustic sleight of hand? The proximate answer was 17,000 square feet of sound absorption that had been artfully camouflaged within the atrium's design by the acoustics firm Acentech in Cambridge, Massachusetts. The thin steel beams supporting the glass curtain walls were faced with a black-metal grill that hid a thick layer of fiberglass insulation. Overhead, translucent vinyl had been stretched across a steel frame that hung from the ceiling; perforated with tiny invisible holes, it sandwiched a 9-inch air gap that trapped sound. Finally, behind the old facade's brick balustrade stood a series of thin wooden fins that each contained a core of sound-swallowing insulation.

These clever sound-dampening treatments had made the atrium somewhat more costly to build, and yet preventing noise is almost always less expensive than trying to address it after construction is complete. Businesses must close for repairs, replace furniture, open up walls, and reconfigure HVAC systems, along with other invasive measures. Acentech gets plenty of retrofitting work from clients who find themselves overwhelmed by noise. But, according to the firm's president, Ben Markham, "The best and cheapest way to solve a noise problem is not to create it in the first place."

More consequential than the café's specific acoustic tricks, however, may be the design process that spawned them. Specifically, the MFA project was Acentech's first use of a "3D listening room." In such a room, clients are surrounded by speakers and sonically transported into the future building, so they can design by listening.

These 3D listening rooms rely on soundscape models that can re-create the sounds that would emanate from points indoors or outside and bounce around the planned space on their way to a listener's ears; the models also include sound absorption, or barriers, incorporated into the design. These aural simulations are computationally intense. Some of the first full-scale versions, with enough speed and fidelity to inform client design choices, were cobbled together in the mid-1990s by an acoustician named Raj Patel in the New York offices of Arup, a global consulting, design, and engineering firm. Patel realized that most people don't experience and comprehend soundscapes the way acousticians typically describe them—with lots of numbers and terms such as noise-reduction coefficient, speech-transmission index, noise-criteria curve, and sound-transmission class rating. If he wanted to integrate sound meaningfully into building-design conversations, then he would need to somehow make soundscapes just as intuitive to understand as the visual models and colorful renderings used by architects.

In 1996, Patel harnessed enough computational power by looping together several computers to simulate his first soundscape for a client—the project was a concert hall. The speed of auralization rendering increased over the years, so that by 2005, they could be acoustically adjusted in real time. The foundation of these models is a well-known acoustic measure known as the "impulse response," which describes the mix of sounds hitting listeners' ears over time, both directly and indirectly, from echoes bouncing around the room—essentially the acoustic signature of an indoor space created by its architecture, dimensions, and materials. In existing buildings, acousticians determine the impulse response by making loud bangs and then measuring the resulting echoes until the sound eventually fades. For acoustic modeling of a space yet to be built, this foundational metric must be calculated based on architectural plans and then applied like a filter to all the sounds one might hear in the space—from an orchestra in a concert hall to diners in a restaurant.

The soundscapes are then brought to life in three dimensions via speakers surrounding the clients, who can listen to the sonic implications of their design choices. With a few mouse clicks, they can bounce to different parts of a planned space, open and close windows, and conjure the sounds of a large crowd, jet overflights, or rush-hour traffic, while swapping options for window glazing and other building materials, altering HVAC systems, and varying the amounts of sound absorption.

For the MFA project, Markham's team invited key members of the museum's building committee into their 3D listening room, where they sat at a table with place settings to enhance the simulation of gathering for lunch in the future atrium café. Suddenly, the surrounding speakers immersed the gathering in the noise that would be generated by five hundred diners in the space as originally designed, with no special acoustic treatments.

"It was total cacophony. You could tell right away that you couldn't talk during your meal or you'd end up hoarse," Markham recalled. So his team started adding sound absorption options to the design in multiple configurations and quantities, letting the museum's project team listen to the results and then making further adjustments before listening again.

"We finally got to a point where it was still lively, still exciting, you still had five hundred people around you, but you could also have a conversation without shouting," he said.

More than a decade later, 3D listening is becoming increasingly common in building design, but it's still a specialized tool used mostly by high-end acoustics firms for prestige projects. Markham expects design by listening to rapidly go mainstream as more powerful computing meets the growing trend of building mixed-use developments, which plunk music venues next to offices and locate bars and restaurants below residences, to realize a vision of walkable communities, with amenities close at hand. But this convenience comes with risk: a great deal of potential noise near people who are trying to work, study, or sleep. Developers of these communities will put a premium on creative soundscape planning.

"My answer is not to say, no, we can't live, work, and play in the same place," said Markham. "I say bring it on. Let's figure out how we can make it work."

Healthier Buildings

Beyond the sonic puzzles presented by mixed-use developments, a deeper trend is bringing discussions of sound into mainstream architecture and building design. A nascent healthy buildings movement, which took off during the early years of the Covid pandemic, has continued growing as companies try to lure employees back to the office.

The Healthy Buildings program at Harvard University, founded in 2014 by an exposure scientist named Joe Allen, stipulates that a building should have ample ventilation and clean water, and no toxic chemicals should be wafting up from its carpets and furniture. It should also provide adequate lighting, thermal comfort, and sound environments that neither endanger hearing nor pile on stress.

The son of a former New York City homicide detective turned private investigator, Allen worked for his dad as a young man before earning his doctorate in exposure assessment and environmental epidemiology. He then spent six years solving sick-building mysteries, including outbreaks of Legionnaires' disease in hospitals and cancer clusters in commercial offices. Allen found the work fulfilling but also frustrating. His corporate clients routinely neglected the modest upfront investments that could have avoided the costly health crises he was hired to fix.

This short-term thinking was partly due to a lack of awareness. Environmental health science tends to be hidden away in academic journals, and its findings are heavy on jargon and light on real-world recommendations. So, after Allen joined Harvard's faculty, one of his first projects involved collaborating with building owners, facilities managers, and real estate professionals to boil down the research into something more straightforward: "The 9 Foundations of a Healthy Building." Noise is included, along with water quality, dust and pests, ventilation, air quality, thermal comfort, lighting and views, moisture, and safety. Each foundation features a two-page research summary of health risks and general guidelines for prevention and maintenance.

The noise pages, for instance, call for protecting occupants from sound disturbances such as traffic, while isolating the buzzes and rattles of machinery and office equipment, providing quiet areas below 35 decibels, and capping room reverberation time at 0.7 seconds (longer reverberation times, in which sound carries and lingers, can be good for some indoor spaces such as auditoriums and concert halls).

Acoustics are also part of several "healthy building" certifications, such as Fitwel, the Living Building Challenge, and the WELL building standard, which have sprung up in recent decades. They give the trend some currency in the real estate market. A 2021 analysis by MIT researchers, for instance, found that buildings with healthy building certifications in ten American cities commanded rents up to 8 percent higher than those of comparable noncertified properties.

Every healthy building certification scheme has champions as well as critics, who say it is either too complicated and expensive for most developers, or too reliant on checklists rather than actual building performance, or too aligned with industry partners making standards-compliant products. Allen remains agnostic about these certifications, but he acknowledges their power to promote healthier buildings as a holistic agenda rather than a set of siloed concerns. For example, a workplace with good ventilation but stagnant water in dead-end plumbing that's brewing pathogens isn't healthy, and neither is an office with superb air quality but terrible acoustics causing chronic stress.

"I think expectations for buildings have shifted," Allen said. "At least on the business side, a healthy building is not just seen as a nice to have, but more like a must have."

The pandemic forced people into work-from-home arrangements that quite a few weren't eager to abandon after the crisis subsided. Not only did the fear of contagion linger, but many people found they could work quite comfortably from home* and spare themselves the commute, the office politics, and yes, the noise distractions.

The magnitude of these changes forced a complete rethink about what we want from our shared workplaces, including how they should sound. At the very least, we want to work in buildings that don't harm our health—via pathogens, chemical pollutants, dangerous decibels, or chronic stress. But could we aim higher? Could we do more than avoid noise and cultivate soundscapes that support employees' focus and improve their well-being?

That's the ambitious goal behind nature-inspired biophilic design, which is mentioned in Allen's nine foundations (but only in the visual sense, under "views," where most of the research has focused). In fact, the only sounds used in his lab's biophilic design studies have been loud traffic and machinery noise that heap stress on people already doing tricky mental math in a virtual reality office. The researchers then tested whether views of real nature through an office window or bits of nature brought indoors, such as potted plants and fish tanks, were better at helping people recover from the ordeal.

While it's well known that spending time in nature is good for us, we've barely started to dissect the sensory components of its salubrious effects. How much benefit is derived simply from the exercise of a long walk in the woods, for example, as opposed to the immersion in greenery or the antimicrobial chemical com-

---

* Not everybody, it should be stressed. Plenty of people worked jobs that couldn't be done from a distance, or they lacked the space and a reliable internet connection to efficiently work from home.

pounds emitted by trees (called phytoncides) that we breathe in with the fresh air? How much does it matter if we can hear the birds and a gurgling brook as we walk, or if those sounds are drowned out by the noise of a nearby highway? Such questions get even more complicated when "nature" is brought indoors.

In one study, for example, Swedish scientists found that nature sounds were vital to stress recovery in a virtual reality forest. They had study subjects give impromptu speeches and do rapid-fire arithmetic before a stone-faced virtual audience before beaming them into various environments to relax. Observing cardiovascular indicators and salivary cortisol levels, the researchers found that a silent forest was no better than a boring nondescript waiting room in helping subjects recover from stress.

Meanwhile, only a handful of studies have explored the use of nature sounds for masking office noise and reducing workers' stress. In 2019, about three dozen employees of the Mayo Clinic in Rochester, Minnesota, volunteered to relocate their 9-to-5 gigs to the nearby Well Living Lab, which had been configured as office space. Over the course of eight weeks, the researchers cycled the office through a baseline setup and three biophilic conditions—visual (indoor palms, flowering plants, and artwork depicting natural scenes), auditory (piping in natural sounds), and multisensory (both plants and natural sounds, plus a small indoor water fountain).

Workers wore sensors that monitored their stress levels via heart rate variability and fingertip skin conductance. They also responded to frequent short surveys about their mood, stress, and job satisfaction. Finally, they took biweekly tests to gauge their working memory and attention.

In September 2021, the researchers published a mixed bag of results in the *Journal of Environmental Psychology*. Natural sounds were the best at improving working memory and other cognitive functions, but only the visual and multisensory biophilic conditions reduced workers' stress. In addition, while workers had an easier time switching their attention from one task to another in an office filled with nature sounds, they performed *worse* than baseline when plants were added to the biophilic mix, perhaps due to sensory overload.

The bottom line is that listening to nature is no magic elixir, especially when its sounds are transported into an indoor environment, where they are obviously simulated and our focus is otherwise engaged by the daily grind. After all, one of the best parts of spending time in real nature is the chance to untether our attention from everyday worries; for most of the workday, such a mental escape is impossible. Adding potted plants to an office or opting for natural materials over particleboard is one thing, but superimposing a seaside soundscape onto a high-pressure workspace is quite another. As I would learn from a leading soundscape practitioner, getting it wrong can backfire, and getting it right takes a bit of science and a lot of trial and error.

Nature Incognito

The first thing I learned from Evan Benway, the founder and chief executive officer of Moodsonic, a London-based startup that creates biophilic soundscapes, was that you have to be careful with birdsong.

Out in real nature, the chirping of birds is a quintessential sound loved by many as well as a potent stress reliever and mood enhancer, as noted in chapter five. Yet birds are also likely to snag a person's attention—that's great when you're on a hike and immersed in a woodsy reverie but not so terrific when you're analyzing a complicated spreadsheet or sweating the final details of a lengthy negotiation.

"We want to support attention, not catch it," Benway explained. "Birdsong is really polarizing. It can drive some people nuts." Several other nature sounds must be handled with care when transplanted to a work environment. The nocturnal sounds of crickets and cicadas, for instance, shouldn't be heard at high noon. Similarly, the crackle of fire may be brain candy around a campsite, but it can be unsettling or even cause alarm inside a building. The same goes for certain water sounds, which might suggest leaking pipes.

Navigating human attention is at the core of Moodsonic's business and remains its biggest challenge, Benway explained, when we met in the London headquarters of Vaimo, a strategy and design firm focused on digital commerce and one of Moodsonic's clients. Moodsonic's algorithms change the sonic mix throughout the day, sprinkling in some randomness to prevent predictability. The human mind excels at recognizing patterns, which Benway discovered the hard way, years before Moodsonic existed, in his earliest experiments with biophilic soundscapes. Even though the nature sounds he tested in one office played on *very* long loops, not hourly or daily ones but weekly, the employees still caught on and came to despise the repeated chirps. "They'd say, 'It's Thursday, 9 a.m., and I hear that damn seagull again!'" Benway recalled.

At Vaimo, clusters of workstations are spread across the hardwood floors of a renovated warehouse, creating an expansive open space with oversized windows, black steel beams, and exposed ductwork. In short, it is exactly the sort of place where halfalogues—one side of phone conversations—can linger and noise distractions proliferate. Before activating the soundscape, Benway introduced me to what inspired it. He showed me a map of Rottnest Island, located about 12 miles from Perth on the west coast of Australia. Visitors flock to Rottnest for white-sand beaches graced by the crystal-blue shallows of the Indian Ocean. They bike the island's perimeter and take selfies with quokkas, cat-sized marsupials with round cheeks and protruding front teeth that seem to be cast in a perpetual grin.

After Benway tapped the screen, a gentle lapping of waves could be heard, the start of a sonic journey around Rottnest, which Moodsonic's algorithms generate fresh each day, based on recordings from the island's beaches, woodlands, swamps, and heaths. A tranquil mix of sounds—the ocean, wind, and gurgling streams, even a smattering of frogs croaking and insects buzzing—flowed from a host of small black speakers camouflaged by the office's steel pillars.

While the soundscapes were infused with randomness, they could be tuned to suit different areas of the office and changed subtly from place to place. In the absence of walls, these sonic shifts helped define zones for specific types of work. A livelier, more engaging soundscape might fit a common area designated for socializing or collaboration, while a low-key blend of sound masking drawn from a babbling brook might better support focused work.

Vaimo's soundscape can also adapt to acoustical changes in the office. Benway demonstrated this by playing a recorded voice on a nearby Bluetooth speaker, which was meant to simulate a talkative coworker. As he boosted the voice's volume and the words became more intelligible, the medley of seaside masking sounds grew louder in turn, but only up to a point. The goal wasn't an acoustic arms race. In fact, Moodsonic had turned down work from clients whose offices had baseline problems with loudness. "We say, you have to mitigate that problem first," Benway told me. "We don't want to simply pile sound on top of sound." On the day I visited Vaimo in London, the place was nearly empty, thanks to a record-breaking heat wave, which had buckled pavements, warped train tracks, and ignited grass fires in the city's outskirts. Local authorities had advised people to stay home.

Even without the crippling heat emergency, Vaimo's office was rarely packed the way it had been before Covid, back in 2019, when the expanding company was outgrowing its headquarters located elsewhere in London. According to Stephen Hill, Vaimo's head of people and culture, the relocation prompted company executives to finally tackle a problem with office noise that had become the top grievance listed in monthly surveys of employees' satisfaction and well-being. During my visit, Hill shared some of the anonymous responses he'd compiled from those surveys:

The thing that would make Vaimo better: Somewhere quiet.

My workplace has spaces to chat and relax, but NOT to focus and work :(

Same as previous months, and what everyone else says—too much noise

On and on went the complaints about noise, filling up the spreadsheet Hill had used to help make a business case for attending to the company's soundscape. He said Vaimo's less cramped post-pandemic office environment hadn't changed those calculations. Stray voices and noises can be even more distracting in a somewhat emptier workplace. Plus, the purpose of biophilic soundscapes isn't simply to "cover up" the office hubbub but to replace it with something more restorative and supportive.

Hill felt that an improved soundscape at Vaimo might be a perk to attract and keep talent. Other companies, in the Covid pandemic's wake, had tried to coax people back to the office with nature-adjacent offerings, such as rooftop vegetable gardens and "treehouse lounges" brimming with greenery. At Vaimo, the nature fix would be delivered by ear.

In Vaimo's small conference room, a gentle acoustic guitar melody drifted down from speakers hidden in an overhead light fixture. This was one of the few Moodsonic soundscapes with musical elements, which are risky because musical tastes are varied and deeply personal. This room was also one of the only places at Vaimo where employees had direct control over soundscape options, including a handy off switch.

"User control is a big, interesting topic for us," Benway said. "Generally speaking, if you can give people more control over their space, then they should be happier. But employers don't always want that."

In a smaller enclosed place like the conference room, direct control was both warranted and unlikely to disrupt the rest of the office. But in a larger shared space, giving everyone the ability to swap out the gentle rise and fall of waves for a raging waterfall, or to shut off the forest on a whim, could be jarring. At the very least, Benway said, it would draw attention to the soundscape, something Moodsonic tries assiduously to avoid.

With or without birdsong, biophilic soundscapes are not the acoustic answer for every workplace. As noted earlier, there is no soundscape perfectly attuned to every work situation. There are no simple rules to follow, like *just add water sounds* or *quieter is always better*. The key is to be proactive about the potential of sound and to expand the range of its possibilities. Instead of worrying only about sounds that exceed certain decibel thresholds, we should design by listening, and by considering an entire soundscape. Whether the goal is creating a lively buzz in a restaurant where diners can still converse without shouting, an office that supports both collaboration and focused work, or buildings that sustain health rather than challenge it, the *right* soundscape always requires a balancing act.

9

BEYOND QUIET

HEARING THE FUTURE CITY

Imagine you're taking an early morning run through your local park, meeting a friend for lunch in a bustling city square, or indulging in a head-clearing stroll around your block at the end of a busy day. What do you hear? What are the key sonic ingredients of these experiences?

Traditional noise control has little patience for such inquiries—its only concern is with the sounds we *don't* want, and its sole soundscape ambition is quiet. Yet we are not walking, talking decibel meters. Our perception of noise is elastic, swayed by our attention as well as innate and learned associations with certain sounds. The same goes for our sense of quiet and the many other sonic possibilities that have long been ignored.

When we think about improving soundscapes rather than simply controlling decibels, sounds become our potential allies— judged not by their loudness alone but by how they shape our experiences. What if urban planners pushed beyond decibel-contour maps and pursued a wider variety of soundscape goals—such as steering the development of a future public square so that its soundscape is vibrant rather than chaotic? Or nudging an urban green space so that it creates a sense of calm rather than monotony?

Such ambitions face massive challenges. When we venture beyond the decibel, we encounter murky, subjective, and shifting sonic terrain. City builders will need new metrics and tools if they are to consider entire soundscapes in places where decibels now hold sway. Bridging that gap will require new insights into sensory experience and the ability to plumb subjective perception to find common denominators that can serve as guides to sonic environments still uncharted. But if these challenges can be met, city sounds can be harnessed to help us attain larger goals for the places we create.

New Ways Toward Quiet

Ask anybody who is bothered by noise what they want, and they'll most likely say quiet—of course. Our world needs more quiet, both as a refuge and a state of mind. But what actually makes a place "quiet"? What does quiet offer us? Why do we value it? And can we find the answers to these questions only by measuring decibels?

Such inquiries are more than academic for Simon Jennings, the executive scientist for the city and county council of Limerick, Ireland. Jennings oversees the five-year noise action plans that the European Union has required from cities for more than two decades, and Limerick's population recently topped 100,000 people, the point at which noise plans are expected to designate "quiet areas," places with a daily average loudness below 55 decibels,* that are to be protected against further noise encroachment.

Limerick is a compact city that hugs the River Shannon before it widens into an estuary and empties into the Atlantic. The banks of the Shannon and its tributaries are popular places for people to meet, meander, eat, and exercise, including Jennings, who can be found at the riverside most evenings, walking his springer spaniel, Trixie. "Those walks are very relaxing," he told me. "The sound of the river is almost meditative."

Jennings went on to describe these riverside strolls as tranquil and peaceful but definitely *not* quiet, at least not in the decibel sense. In fact, the Shannon can be quite loud as it tumbles through weirs and over rocky areas. At some stretches, the rushing river is all a person can hear, even where the path edges close to a busy roadway.

While bigger European cities harbor very-low-decibel areas deep within their largest parks or on the city's outskirts, Limerick's parks are small, making it harder to escape the usual hubbub of urban life. For years, Jennings had measured sound levels throughout Limerick and studied the city's decibel-contour maps. So he knew that Limerick's most restorative places, such as the riverside paths or the shady "pocket parks" tucked away from busy streets, would not likely qualify as quiet according to the accepted decibel threshold.

As the city's noise chief, Jennings had nothing against low

---

* By the letter of the law in the 2002 Environmental Noise Directive, the measure of quiet could be anything. But the only metric actually used is decibels, and most cities have defaulted to a threshold of 55 Lden (a twenty-four-hour average, with a 5-decibel penalty for noise between 5 and 11 p.m. and a 10-decibel penalty for overnight noise).

decibel counts. He simply felt that this single measure was a poor proxy for something far more meaningful and worthy of protection, if only it could be measured and mapped.

He wasn't alone. In 2019, Jennings started working with an Italian architect and urbanist named Antonella Radicchi, who had set out to redefine quiet with the help of a crowdsourced app called Hush City. Radicchi, now an independent researcher and consultant with the European Urban Initiative, is part of a growing movement to challenge decibel levels as the sole arbiter of good versus bad city sounds. Before becoming an academic, Radicchi had worked as an architect but felt her field had grown too visual, forgetting that people experience places with *all* their senses. When she joined the Technische Universität Berlin in 2016 as a postdoctoral research fellow in urban planning, Radicchi decided to focus on combatting urban noise.

Berlin's authorities had already mapped out several very-low-decibel areas in their five-year noise plans. But these places were mostly situated deep within huge parks and were therefore not easily accessed by busy Berliners. If the goal of protecting quiet was to bolster public health and well-being, Radicchi argued, then cities should cultivate more "everyday quiet areas" where residents could easily find a peaceful refuge within walking distance of their home and workplace. There, they could escape the din, read a book, chat with a friend, or simply sit in peace for a few minutes to relax and decompress.

Radicchi was not simply looking to loosen the decibel threshold of "quiet" to better accommodate boisterous urban realities. She wanted a new definition of quiet areas that pushed beyond the decibel counts, which she felt missed something crucial about what made these places valuable and restorative for city dwellers.

Emerging research backed her up. For instance, it is possible to make noisy places seem quieter to visitors by *adding* sound— the right sounds in the right ways. Numerous interventions have proved this an intriguing idea. For instance, researchers in the Sounds in the City Lab at McGill University in Montreal, which works with city officials and building professionals to improve urban soundscapes, transformed the vacant site of a former gas station in a dense commercial district into a public square called Fleur de Macadam (Flower of the Asphalt), surrounded on three sides by vehicle traffic. In the summer of 2018, they collaborated with local officials and their contractors to create different temporary versions of this space by arranging seating, tables, planter boxes, and lighting, and testing two distinct sound installations played from speakers perched on lighting poles. One consisted of a sparse mix of nature sounds with percussive and melodic musical elements, and the other included snippets of voices speaking over a background of music and typical urban sounds.

Objectively speaking, the sound compositions added more sound to an already noisy setting, and yet hundreds of surveyed visitors rated both sonically enhanced versions of the square as calmer, more pleasant, and *less loud* than the place seemed during a control setup when the speakers were turned off. The McGill team replicated these results the following summer, despite the added challenge of loud roadwork going on nearby.

Likewise, researchers in Singapore found that people rated traffic noise as less loud when it was augmented with birdsong or the gurgle of a flowing stream. In a follow-up study, they showed that people in an outdoor seating area near an expressway perceived the traffic as less loud and gave the quality of the soundscape higher marks when birdsong or fountain sounds were added via a loudspeaker or an augmented reality headset.

Back in Berlin, Antonella Radicchi knew that top-down, decibel-defined quiet doesn't take into account the complexity of human perception. So she decided to redefine quiet from the bottom up by crowdsourcing the effort. No matter what a decibel meter said, she figured Berliners knew a restorative place when they experienced one. Why not start by asking them for their insights? To that end, Radicchi created the Hush City app to gather and compile information on the everyday quiet areas identified by users, which then appeared on the app's searchable maps.

"I wanted people to share their own quiet areas and evaluate them independent of me and experts in general," she recalled.

Hush City does have a decibel meter, but that provides only one data point among many. Submissions also include photos of a place's scenery and answers to a short survey describing the variety of sounds that can be heard there, and which elements enhanced or detracted from its sense of quiet; these items are rated on a five-point scale. The app also provides a menu of adjectives to capture the overall feel of the place, such as "natural," "lively," "familiar," "boring," "stressful," or "relaxing." Most people chose several descriptors, but the app highlights the first one they pick, along with the locale's decibel count and a photo to help people looking for nearby quiet.

Between April 2017 and February 2018, Radicchi piloted Hush City in a rapidly gentrifying Berlin neighborhood called Reuterkiez, which had relatively high levels of air, water, and noise pollution, as well as limited green space, according to the Berlin Environmental Justice Atlas. What did quiet mean to these folks?

As one might expect, ratings of quiet rose where decibels dipped and greenery grew. The most popular "everyday quiet areas" in Reuterkiez were two small parks, each about one square block in size, as well as walking paths along a canal on the neighborhood's northern edge. But there were also some surprises. For one thing, people identified a wide variety of quiet areas in small playgrounds, hidden gardens, outdoor cafés, and even the sidewalks of streets with minimal traffic. Also, app users seemed to value places that favored social interaction and conversation—while "relaxed" was the most popular descriptor for these everyday quiet places, "lively" came in second.

By the end of Radicchi's pilot of the app, word had spread about it, and she had more than five hundred submissions to analyze from all over Europe, the United States, and Australia.* Quiet ratings were boosted by certain sounds, namely, birdsong and human voices, while other sounds, traffic in particular, were especially unwelcome. Vehicle noise was detrimental to the sense of quiet 55 percent of the time, far more than the runner-up annoyances, such as noise from airplanes and HVAC systems, which were blamed for disrupting quiet from 1 to 3 percent of the time.

Meanwhile, feelings of security rose along with quiet ratings, but only up to a point. In "very quiet" places, that sense of security diminished. In fact, Hush City is chock full of very-low-decibel spots that people found *boring, ugly,* and even *stressful,* while some higher-decibel areas—usually topping out around 70 decibels—were deemed *relaxed* or *natural.* As Radicchi put it, "People in cities don't look for silent places. They look for places with high sonic quality."

Later in 2018, authorities in Berlin partnered with Radicchi

---

* The Hush City app featured more than 7,300 everyday quiet areas as of this writing.

on a citywide expansion of the Hush City project. The result was Annex 10 of the city's 2019–23 noise plan, *Berlin Wird Leise* (Berlin Is Getting Quiet), which recommended that the city find a way to officially recognize everyday quiet areas.

"Small-scale retreats contribute significantly to a sense of well-being in an increasingly dense city," the report stipulated. "They are important for protecting the health of the population because they offer opportunities for relief from the everyday noise situation." Rather than meeting a low-decibel cutoff, this new category of urban rest stop should offer a "clear contrast in acoustics and design" to the surrounding cityscape. However, more specific criteria for everyday quiet areas remained to be defined; the authors of the report suggested that multiple city departments would need to weigh in on the matter in the coming years.

Back in Limerick, Jennings wanted his city to follow a similar path. He convinced the city and county council to OK the Hush City app to crowdsource candidates for quiet places, but the pandemic interrupted the effort. Unwelcome quiet descended everywhere, and soundwalks (excursions focused on listening to different places) were canceled. When Jennings and I first spoke in 2022, his search for everyday quiet areas was barely sputtering back to life.

"It's not been an easy process of engaging the public," Jennings admitted.

By early 2024, Jennings was preparing Limerick's next five-year noise plans (one for the city and one for the county), which would propose several public green spaces as quiet areas, even though he suspected they would ultimately exceed the 55-decibel threshold. To that end, he was also planning a campaign of soundwalks and citizen surveys, combined with extensive acoustic measuring, to rank the green spaces and bolster the case for designating them as quiet areas.

Meanwhile, Jennings had been mulling a related idea. If everyday quiet was about more than decibel levels, then maybe efforts to identify and protect it should expand beyond the city's noise-control plans. He thought he might generate more engagement by organizing soundwalks with schools, local environmental groups, and mental health advocates, for instance—people who were less interested in counting decibels and more interested in enhancing the city's green space and help people relieve stress and anxiety.

Tuning the World

Efforts to reexamine quiet in Limerick, Berlin, and elsewhere raise critical questions: What *is* the purpose of quiet places? Are they simply a bulwark against the spread of urban noise, like a dam holding back a rising tide of loudness? Or do these places have their own sonic identity and a distinct role to play as building blocks of a healthier and more restorative city?

The former rationale puts quiet areas squarely in the noise-control column. The latter suggests a variety of sonic goals that cannot be assessed by decibel counts alone but by measures of "high sonic quality." This more expansive approach now counts one of the world's leading noise fighters among its biggest boosters.

Lisa Lavia became the managing director of the UK Noise Abatement Society in 2009. A native Californian who moved to England in the early 1990s, Lavia is a self-described noise-sensitive person who has become a soundscape evangelist. On occasion, she has even lobbied for more sound rather than less in her adopted hometown of Brighton, on England's southeast coast. She once convinced Brighton's city council to try *adding* sound to the heart an infamously rowdy clubbing district, where drunks caroused in the wee hours, brawling, shouting, and drawing a heavy police response as well as an overwhelming number of noise complaints.

Lavia's partner on the project was Harry Witchel, a neuroscientist and body language expert at the Brighton and Sussex Medical School. From surveys and community meetings, they knew that nearby residents generally liked living in a vibrant part of the city, were attracted to the neighborhood's energy, and didn't mind some weekend revelry. What people did mind was chaos, violence, and feeling unsafe. So, when Lavia and Witchel suggested that they might help subdue the mayhem by imposing a different sort of sound mix on the clubbing district's main drag, West Street, residents were at least willing to let them try.

They partnered with the artistic director of Brighton's White Night Festival, Donna Close, to commission a pilot sound installation from Martyn Ware, formerly of the 1980s pop band The Human League, who has since become a well-known sound artist. On a Saturday night during the festival, Ware's composition flowed from well-camouflaged speakers on either side of West Street, immersing its drunken denizens in a soothing mix of natural seascape sounds, ethereal synthesizer, and samples from pop hits such as Beyoncé's "Countdown" played at half speed. The vibe of the area changed dramatically, replacing the usual angry shouts, shattered bottles, car horns, and police sirens with something closer to groovy mood music. Compared to the typical Saturday night experience, the number of noise complaints plummeted, and police were able to keep a lower profile, even redeploying units to other parts of the city.

For Lavia, the lesson of the West Street project was not that noise problems should be solved with bespoke pieces of sound art. Indeed, a lack of funding meant that Ware's installation could not be made permanent. Lavia had a different takeaway: avoid preconceived notions about what a soundscape ought to be and what approaches to noise issues should look like.

In the summer of 2022, she took me on a quick soundscape tour of Brighton, starting with Palmeira Lawns, a narrow Victorian greenway crisscrossed by paths, benches, trees, and flowering shrubs. It was flanked by a long block of white Italianate residences stretching to the shore. We stood at the park's northern end, which borders a busy road where a bus rumbled to a stop and coughed up a guttural idle.

"One principle of soundscape is that there are no bad sounds. There are simply the wrong sounds at the wrong time," Lavia said, smiling as she raised her voice over the buzz of a passing motorcycle's two-stroke engine. She gestured toward the paths radiating through flowerbeds and a grassy expanse, where a couple of dog walkers and their tethered companions exchanged quick greetings.

"This is what I'd call very high-quality green space. But I could never relax here," she said. "And that's a challenge for a lot of city parks. If noise pollution isn't taken into account, then those parks may not be doing their job."

The bus throttled back to life, then departed. Lavia grimaced and led the way down a path toward the park's southern end, where the noise dropped dramatically. By the time we reached a large concrete birdbath encircled by a rose garden, the trees, plantings, and gently rolling topography had swallowed most of the traffic's roar. In this more tranquil zone, people relaxed on benches to eat lunch or read books. A woman in running gear stretched out on a patch of grass, gradually unfolding herself until she lay there, supine and still. Other sounds reached us here. A dog barked cheerfully. A gull's call announced its long glide to the beach.

"These little hills and vegetation have effectively created a noise barrier," Lavia explained. "It's not necessarily done intentionally. But it *can* be done intentionally. The problem is that architects and urban designers aren't trained in sound."

From the base of Palmeira Lawns, we descended some stairs and crossed a busy coastal highway to reach the waterfront. As we walked along a promenade between the pebble beach and a string of cafés, boat pavilions, and fish-and-chips stands, Lavia recounted the story of her soundscape awakening.

Shortly after taking charge of the Noise Abatement Society, she embarked on a listening tour, meeting with acousticians, government authorities, and the public to help select the society's next target. Rather than revealing her newest noise enemy, however, these conversations convinced Lavia that she'd been missing a much bigger picture.

"Everybody knew what the problems were. People knew noise pollution was harming health. The government wasn't denying it. Industries and businesses were spending a lot of money trying to fix these problems," she explained. "If we just campaigned to stop one more source of noise, then all we'd be doing is telling people what they already knew."

At the same time, Lavia had started learning about sound-scapes and the importance of listeners' perspectives and contexts, including a handful of collaborations between academics, sound artists, and community leaders that had revitalized neglected public spaces with the help of soundscape design. Lavia was drawn to these ideas, but she knew it would be a struggle for them to get a foothold in the workaday world of most building professionals. From building codes to project bids, any consideration of sound was based on standards.

"Architects weren't going to change how they design buildings, and city planners weren't going to disrupt their practices simply because somebody said it was a good idea," Lavia said. She had found her noise challenge. Making common cause with others trying to close the gap between soundscape theory and building professionals' practice, Lavia dove into the daunting task of creating standards for a field that maintained a healthy skepticism of anything standardized.

She joined the International Organization for Standardization's newly formed committee on soundscapes in 2010. It took nearly a decade of negotiations, but eventually the ISO issued a definition of soundscapes and guidance for assessing them, using a mix of acoustic measurements, recordings, and listener feedback gathered from questionnaires, soundwalks, and interviews. The main premise: perception of a soundscape is subjective but not arbitrary. Put another way, no two people will experience any soundscape in the same way, but on a larger scale, there are meaningful trends and clusters of reactions to any sound environment.

In 2018, the committee published a how-to for plotting the entire universe of soundscape perception on a simple graph with X-Y coordinates. It was based on work by Swedish environmental psychologists who had boiled down every typical descriptor of a sound environment—from "tranquil" to "fun" to "eerie" to "cacophonous"—to an intersection of *pleasantness* (on the X axis) and *eventfulness* (on the Y axis). They added measures that cut diagonally across the quadrants, such as *exciting* in the upper right (eventful and pleasant) to *monotonous* in the lower left (uneventful and unpleasant). The result was an eight-spoked gauge of perception called the soundscape circumplex, which looks like this.*

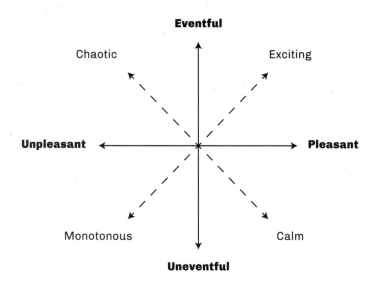

Decibel levels were not irrelevant to these measures, but they didn't dictate them nor determine the type of soundscape desired. Sometimes people wanted to immerse themselves in the energy of a lively city, or enjoy the familiar hubbub of their neighborhood, just as they sometimes craved periods of calm reflection.

---

* The ISO version of the graph included a couple of tweaks to the wording. "Vibrant" replaced "exciting," and "annoying" replaced "unpleasant," for instance. In addition, a Z axis, for *familiarity*, has since been proposed and debated by soundscape scholars.

From a Sonic Sketchpad
to Predicting Perception

But the job was not done. Standards nobody uses are dead letters. Lavia and other backers of the soundscape approach still needed real-world city builders to give their ideas life. For urban designers to take such a risk, they would need tools to reliably predict reactions to sonic environments that did not yet exist, to generate excitement without tipping into noisy chaos, for example, or to keep things calm without boring everybody's ears off.

Different ideas for these sorts of soundscape tools have begun to emerge. One is similar to the immersive "design by listening" approach to planning indoor acoustics. McGill's Sounds in the City team, for instance, is developing a virtual reality tool that urban planners could use to upload digital mockups of a new city square or shopping district and then add, subtract, and rearrange sources of sound, ranging from bus traffic on a city block to water features in a park.

By late 2023, the McGill project was being led by a doctoral student, Richard Yanaky, who created a desktop prototype, which he called a "sonic sketchpad" for city planners. It includes an extensive menu of sound interventions that could be embedded into six square blocks of a virtual city, with buildings and traffic patterns that can be modified. There was road construction on a side street, and the public square featured both green space and a pavilion big enough for a café or an outdoor music performance. Noise mitigations included detailed ordinances, such as limiting musical performances or construction work to specific hours. An adjustable time slider tracked the soundscape as it changed through the day, evening, and night.

Still under development, the sonic sketchpad is being readied for use in a course on sound in urban design that the McGill group plans to offer to architects and planners. "We might ask them to create a relaxing atmosphere, or one that would be good for shopping or a night out on the town," Yanaky said. Eventually, developers will be able to upload their own computer-aided designs and the sources of sound pertinent to a specific project into the sketchpad's open-source software, Yanaky noted. He hoped the desktop version would also be upgraded to a more immersive virtual reality experience for the brainstorming of soundscape ideas.

"This tool is meant to be an invitation for some rapid soundscape prototyping," he explained. "It's about making some decent soundscape options to pass around to community members and ask, 'What do you think?'"

Meanwhile, a plot to dethrone the decibel has been brewing in London. A cadre of researchers at University College London (UCL),* who specialize in acoustics, urban planning, engineering, and architecture, are working to distill soundscapes into a handful of measurable data points that could be used to predict how future soundscapes will be perceived. They want to empower city builders to compare the calmness of various designs for hospital courtyards, for example, or to shift the

---

* The principal investigator of the soundscape indices project is Jian Kang, UCL professor of acoustics and soundscape. A full overview of the research team can be found here: https://www.ucl.ac.uk/bartlett/environmental -design/research-projects/2023/oct/soundscape-indices-ssid.

sonic mix of a planned development from chaotic to vibrant, using artificial intelligence to help judge the suitability of a future soundscape.

This project, which began in 2018 and has received seven years of funding from the European Union, has collaborators in Italy, Spain, the Netherlands, and China. They began their work by gathering a wide range of data from soundscapes around the world. The London team ventured out to the city's parks, plazas, and thoroughfares to capture sounds and visuals from every direction, using tripod-mounted microphones and 3D cameras. They asked thousands of passersby to rate the soundscapes along the eight dimensions of the circumplex (on a five-point scale). The researchers then used some trigonometry to combine the X and Y coordinates of each person's ratings so they could plot a single summary point of soundscape perception for every individual.

One of the London researchers, a bearded, bespectacled doctoral student named Andrew Mitchell, popped open his laptop and showed off the analysis of data gathered from a tree-lined walkway in Regent's Park. A blue blob enclosed the dots that together summarized people's various perceptions of the soundscape along the path, including all the outliers. The blob grew progressively darker where the responses clustered, and its darkest portion landed on the high end of the pleasantness metric (the X axis) while straddling the line between eventful and uneventful (the Y axis), albeit shaded toward calm in the lower right quadrant. With a click, Mitchell simplified things by dropping all but the darkest blue blob representing the middle 50 percent of respondents.

"To me, that shape is the soundscape," he explained. "You can see that it's entirely pleasant, mostly calm, but with a little bit of vibrancy, which makes sense for where we were in a lovely green area with very little traffic noise, and a little bit of energy from people walking down the park's main path."

That doesn't mean that everybody will experience this part of Regent's Park the same way. In fact, they surely won't. On any given day, a visitor might find these same paths oppressively dull or chaotically overrun with tourists and school field trips. Nevertheless, this blob was meant to show the prevailing perception of the soundscape, without ignoring the outlier responses that tugged at the mean ratings and morphed its overall shape.

Generating these soundscape blobs from visitor feedback was only step one. In order to build a tool that could predict perceptions of sound environments yet to be created, the London team also gathered a ton of data about the soundscapes, so that machine learning could look for patterns linking these bits of data with the perception ratings. They recorded time-stamped snippets of audio and video and took a host of acoustic readings, including decibels, plus psychoacoustic measures such as tonality, sharpness, roughness, and impulsiveness. They also asked people to describe the types of sounds they heard, which sounds were dominant, why they were visiting the place, and so on. Their international collaborators did the same in their own cities.

By the end of 2019, the project team had amassed troves of objective and subjective information about each soundscape, and they used that data to train a computational model that could predict perception. Mitchell envisioned a future in which architects and urban planners didn't simply fall back on decibel thresholds, but fed a few choice acoustic and visual details about a planned site into an artificial intelligence tool to gauge how well the predicted perception of a soundscape fit with design goals.*

If that soundscape fell short, then the tool could point toward possible fixes. Maybe the loudness of traffic noise is pushing the soundscape perception into unpleasant territory, which could be remedied by adding trees or other landscaping barriers to the site plans. Or maybe loudness wasn't the issue, but the HVAC system was annoyingly tonal, something that could be mitigated mechanically or by adding a water feature to mask the unpleasant sound.

When it came to the tool's potential for reading the minds of future listeners, test runs revealed both strengths and weaknesses. Fed chunks of acoustic data about a given soundscape, the model predicted an average perception score that matched the actual measured perception with an impressive 80 percent accuracy. But our perception is influenced by factors well beyond sensory inputs, as the project team learned firsthand in early 2020.

Covid's first wave put an end to in-person surveys and postponed plans for virtual reality experiments based on the 3D video footage.† Nevertheless, the researchers still managed

---

* By 2024, the team had devised a framework for scoring this fit of urban design goals and the predicted perception of a soundscape, which they called the Soundscape Perception Index. For specifics, see https://osf.io/preprints/psyarxiv/zpa9e.

† This virtual reality piece of the project was only starting to get underway again at the time of my visit. One major effect of the pandemic-related delay was that the prototype perception model had yet to incorporate any visual information.

to gather data from soundscapes, which now averaged about 6 decibels quieter as traffic thinned and people kept a nervous distance from one another during lockdowns. When the team plugged these new inputs into the pre-pandemic perception model, it yielded some interesting results.

As Mitchell explained it, the model essentially reflected how people in the pre-pandemic world, people who knew nothing about the virus and the mayhem it unleashed, might have perceived the newly hushed soundscapes. For instance, Venice's Piazza San Marco, normally awash in noise from tourists and the businesses catering to them, was a ghost town, empty and shuttered, during the first phase of the pandemic. Even without this specific, tragic context, Mitchell said, "It's just really eerie to suddenly go from a carnival atmosphere to deserted." Yet the model lacked this human understanding, and it predicted an upsurge in perceived pleasantness and calmness for the suddenly barren piazza.

These results raised critical questions: How "normal" was the perceptual mapping from 2019? How much would it have differed from, say, 2010, 1990, or 1950? Maybe (hopefully) 2020 was an outlier, in which extraordinary times jarred otherwise stable patterns of perception. But maybe not.

Either way, our judgment of a soundscape is not only subject to change in different times but also in different spaces. Specifically, we have different expectations of and tolerance for sounds in a busy commercial district as opposed to a leafy park. At the very least, Mitchell suggested that future tools will need to be calibrated to predict perception in different *types* of places, such as large urban squares and residential neighborhoods.

Finally, broad cultural forces influence soundscape perception. For instance, the project's Chinese collaborators needed three or four characters to *almost* match the English concept of "vibrant," and China's city parks are famous for exuberant, high-decibel performances by musicians, dancers, and karaoke singers. That said, cultural norms are constantly in flux. Chinese cities roiled with controversy in 2021 when new noise pollution laws set severe limits on the so-called dancing grannies. These groups of mostly older women, known as "damas" (big mothers), have long gathered in parks and public squares for communal dancing, to music blasting from portable stereos. The new laws restricted when the grannies could meet and the volume of the music they played.

Like the makers of the sonic sketchpad, Mitchell understood the limits of what he and his collaborators could accomplish. There is an inherent entropy to any algorithmic attempt to predict what millions of people will like or dislike about a soundscape that itself is subject to constant change.

"My goal for the eventual tool we hope to release is that it will be just that, a tool," he explained. Periodic additions of new data will be necessary to keep the tool up to date. Human designers will still be needed to infuse the planning of urban places with creativity and inspiration. Most important, for any project, local people must be consulted, hopefully with a growing appreciation of high sonic quality and soundscape possibilities.

The Sense-able City

People who care about soundscapes sometimes say they want to make the world sound better, not merely quieter. It's a convenient shorthand but also something of an eye roller, as if the backers of soundscape design are audio geeks nitpicking the sound quality of a global MP3. The real goal is not angling for some precious notion of sonic perfection but rather raising awareness about the richness of our sound environments and the power they wield over our health and well-being. This influence can't be summed up by a single acoustic dimension.

Soundscapes connect us to places. They are the auditory component of the multisensory urban-design research being done at the Person Environment Activity Research Laboratory (PEARL), a complex the size of an airplane hangar located in a light industrial area on the outskirts of London. The principal investigator, a University College London civil engineer named Nick Tyler, studies how cities can better support human bodies and brains, especially those of older residents or people with disabilities. Subjects in his experiments wear biometric sensors, including skullcaps that track brain activity, as they navigate mockups of a grocery store or a busy urban plaza.

The lab's massive scale lets Tyler make these experiences as true to life as possible while maintaining control over the sensory variables of interest. PEARL opened in 2022, replacing another enormous facility that was apparently not quite enormous enough. When the architects designing PEARL asked Tyler what his vision was, he said he wanted a building that could "house the world." They delivered.

The main experimental space measures roughly 43,000 square feet, and is three stories from floor to ceiling. The interior can be divided by heavy curtains into four bays for simultaneous studies or opened into one massive soundstage big enough to accommodate a line of subway cars, a small park complete with living trees, or several blocks of streetscape. Rockwool insulation nearly a foot thick behind PEARL's walls creates an acoustically dry 30-decibel background, a blank sonic canvas onto which the researchers layer immersive soundscapes through a bevy of speakers controlled from a computer-packed alcove called "the spin." Tyler and his team can also adjust temperature, ventilation, and lighting to simulate bright mornings or dim twilights. They can overlay the thick concrete floor with a range of different surfaces, from wood to stone to turf. There is even a vaporizer system for wafting in various smells, and a fog machine can fill the place with an obscuring mist. Solar panels on the roof generate enough electricity to power both PEARL and several buildings nearby.

Now in his early seventies, Tyler is an avuncular presence with a touch of mad scientist, thanks to a shoulder-length spray of silver hair flowing from his balding pate and the world-creating technology at his fingertips. I met Tyler in PEARL's second-floor reception area, and he warned me to whisper as we ventured into the main lab because an experiment was underway. Researchers were testing options for upgrading the destination signs on buses, comparing the legibility of the existing back-lit roller boards with digital LCD and LED displays, colorized or black and white, as seen from different distances and under various lighting conditions.

"Tomorrow we're going to add fog," Tyler announced as he opened a door and we stepped onto a steel balcony overlooking the vast research space. Off to the right, a double-decker bus faced the study subjects, who stood behind small podiums arranged at staggered distances as they waited for the bus to reveal its next destination. A bit farther off were three long strips of red, green, and blue flooring, each with a wheelchair at one end.

Tyler explained that the City of London—the square mile containing the city's central business district—was hatching plans to enhance pedestrian safety and comfort by blocking off some streets to vehicles, widening sidewalks, and possibly replacing some concrete areas with a softer surface that would ease the wear and tear on the knees and hips of an aging population and hopefully decrease the number of injuries from falls. Nevertheless, these softer surfaces could be tough for people in wheelchairs. The colored strips were surfaces of various hardness that Tyler's team had been testing out with wheelchair users.

Finally, at the far end of the facility was a small parking lot of e-scooters, the focus of the lab's current study related to sound. These quiet battery-powered scooters were proliferating around London but their drivers kept smashing into unsuspecting pedestrians (in 2022, scooter collisions in the city caused hundreds of injuries and three deaths). Like Anderson Consulting, which had worked on alerts for electric buses, Tyler and his team were trying to craft a tone that would keep people safe without flooding the streets with unnecessary noise.

The researchers were still narrowing down the choice of sounds to be played from speakers mounted on scooters as they zipped up from behind study subjects, who would wear skull caps to track brain activity. One should expect to see a spike of activity over the auditory cortex when the brain noticed an alert sound in the wash of urban noise, and then, another spike in the visual cortex when the scooter suddenly flashed into view, Tyler explained. In the simplest terms, his goal was to make that second spike smaller, which would indicate that the sound had successfully alerted the pedestrian's brain to expect a scooter's arrival.

In order to give ample warning, the sound would need to be loud enough to be detected at 10 meters' distance, but no louder than that, he said, "because if you can hear scooters from 20 meters away, then you're going to hear scooters everywhere, and you won't know where to focus."

Measures of sound previously taken from central London suggested that during the day, the city kicked up a background buzz of about 65 decibels. Tyler hoped that by targeting certain frequencies wherein city sounds are relatively subdued, they could tuck the scooter warnings into the larger soundscape without requiring extra decibels to rise above it. When Tyler talked about sound, he spoke expansively about the human need to feel safe and socially connected. The most fundamental purpose of our hearing system is defensive, he explained. "It's so we can detect danger signals from far away and locate them."

Our brains eagerly search for patterns and regularity in sound; those cues help us know what to expect, and that makes us feel grounded and safe. Sound connects us to the familiar rhythms of our environment and to one another, while noise cuts us off and isolates us. To our primal brains, noise is either the strange sound that defies the expected pattern—the proverbial snapping of twigs in the underbrush—or a prevailing auditory confusion that can obscure those same warning signals.

There is no city soundscape, nor even a single sound, that's not embedded in memory and learning and tangled up with our other senses. In our lived experience, the scale of building facades, the controlled chaos of busy sidewalks, the smells of the streets, restaurants, and earthy green parks, and the feel of varied surfaces passing beneath our feet as we move through our days are interconnected. Cognizant of this, Tyler doesn't study city sounds from the narrow perspective of how they might be perceived as noise. His research into sound is just one part of a larger effort: to improve the multisensory experience of city life.

"We can build a hundred-meter streetscape and then manipulate the lighting, smells, and soundscape, and see how you respond to that," Tyler said. "We can combine your behavioral responses with the data we get off wearable sensors to see what your brain's doing, what your heart's doing, how much oxygen you're burning through, and how your stress level changes."

"Our goal," he continued, "should be to figure out how to make the sound environment work with the brain, so that it synergizes with what your eyes are telling you, what your nose is telling you, and what your feet are telling you."

Every aspect of our lives can be enhanced or undermined by sound—from the delicate machinery of sensory perception, with its unceasing and lightning-quick connections to our brains, to the fullness of a city's soundscapes triggering stress or restoration, calmness or confusion, enjoyment or aggravation. In other words, making any piece of our world "sound better" should be part of making it healthier and more resilient, more connected and more welcoming.

CODA

Bose sold the first pairs of headphones with active noise cancellation to consumers in 2000. The global market for such devices has since exploded, set to exceed $45 billion by 2031. These products promise us the power to escape one another's noise by retreating into personal sonic sanctuaries. As the tagline for another noise-canceling product, Beats by Dre, put it, "Hear What You Want." Whether worn over our ears or in them, these devices exude a similar "don't bother me" vibe.

The rise of sonic isolation has coincided with a related trend—people physically together but mentally scattered, glued to their smartphones while sharing a sidewalk or a gathering with friends. In 2019, after Uber gave premium customers the perk of avoiding chatty drivers by selecting "quiet preferred," the *Guardian* published a feature with the headline "Hit the Mute Button: Why Everyone Is Trying to Silence the Outside World."

It highlighted the popularity of "muting" one another on various social media.

"Mute promises a snake-free garden," the writer noted, "a world where you can curate content and silence dissent."

Noise-canceling headphones can certainly be a godsend, allowing us to assert some personal control over our sound-scapes when we really need it. Nevertheless, our ultimate goal shouldn't be a snake-free garden or, for that matter, a noise-free world. Instead, we should listen *more* and think seriously about the sounds of the shared places we want to create.

We have spent a century targeting the world's decibels. But the bigger foe is arguably sonic shortsightedness. Noise has long been dismissed as a problem confined to a few loud industries and private disputes, an issue that's grist for grumps, complainers, and killjoys, and an annoyance disconnected from the broader priorities of public health or environmental protection. Even as technology offers us expanding influence over our acoustic environments, concerns about sound are too often bypassed or sequestered to a decibel count in the design of our buildings and cities. As a result, we endure innumerable cacophonies that might have been avoided.

Meanwhile, the larger fight against noise routinely plays catch-up, trying to roll back sounds after their sources have become inextricably linked with livelihoods, services, and conveniences, rather than anticipating future trouble in an ever-changing soundscape. As Les Blomberg, founder of the Noise Pollution Clearinghouse, pointed out to me, Julia Barnett Rice's war on Hudson River tugboat whistles was contemporaneous with the Ford Motor Company's introduction of the Model T and the maiden flights of the Wright brothers' airplane at Kitty Hawk.

"We have been fighting the last noise battle for a hundred years. I'm trying to fight the next one," he said.

Specifically, Blomberg's worried about a looming sonic scourge—swarms of delivery drones, and eventually, air taxis—that nobody will hear coming until it's too late. Deliveries via drone were still relatively rare when I wrote this book, but their numbers were expanding rapidly. Several drone-delivery start-ups had already launched, and the number of hospitals, restaurants, and established retailers offering drone service was skyrocketing. A 2023 McKinsey report estimated that the global volume of package drops via drone would reach more than a million that year, double the number from only two years earlier.

Proponents of delivery by drone argue that "last mile" shipping by unmanned flying machines will lower costs (fewer drivers to pay) and emissions (fewer miles covered by trucks). Above all other considerations, however, drone delivery will be faster—potentially a lifesaver for people in remote areas who need medicine or other critical supplies, or just another dollop of convenience for suburbanites who forgot to pick up ice cream at the supermarket.

Meanwhile, opponents of drones worry about privacy and safety, but what they viscerally despise is the noise. In 2018, when the drone-delivery company Wing started servicing a suburb of Australia's capital city, Canberra, residents complained that the sounds triggered migraines and symptoms of PTSD for veterans, while disturbing pets and local wildlife. Citizens launched anti-drone campaigns, and some people threatened to shoot the machines out of the sky. This backlash forced the company into a costly redesign of their delivery vehicles and a dispersal of the routes drones flew to and from package warehouses, in order to dilute per-person noise exposure. Some industry observers suspect the fierce opposition played a role in Wing's decision to stop deliveries in Canberra in 2023.

The issue of drone noise follows familiar patterns. Once again, the core of the problem isn't loudness. One analysis calculated that a person standing directly beneath a drone flying 200 feet overhead would endure about 50 decibels of sound from the machine's whirring propellers, or what a decibel chart might label "quiet office." If that person stood 500 feet from the flight path, the loudness would drop to 40 decibels, equivalent to the patter of light rain. Such decibel levels would not come close to violating most municipal noise codes during daylight hours, not to mention the fact that the study measured decibel exposures for a person standing outside, and municipal noise infractions are typically gauged by decibel readings from inside a person's home.

What really bothers people about drones is not their loudness but the tonality of their keening, high-pitched whine, which drills into your brain like the sound of a robot mosquito or a flying leaf-blower. In studies by NASA, people listening to recordings of drones without being told the source found the noise far more annoying than the rumble of delivery trucks and other vehicles played at the same decibel levels.

In the United States, drones flying below 400 feet and outside protected airspace, such as airports, are governed by local rules and regulations. However, few cities have a mechanism, other than decibel levels, for thinking proactively about their soundscapes, and so most municipalities have not seriously considered how to handle a drone-filled future.* Myriad questions remain

---

* Los Angeles is a major exception. In 2021, the city's transportation department published a report on the policy implications of both package delivery by air and air taxis. The department is working with Arup to include noise concerns in coverage of topics such as health and safety, equity, and adaptation of transportation networks, among others.

unanswered. Will drone inspections be required of licensed operators, and if so, will they include noise checks? What are the optimal routes for drones? What hours of operation will be permitted? Where should we put drone depots, which will be hives of busy, noisy activity?

Outside the halls of governance, there's been more active research into some of these questions, including studies sponsored by the drone industry. Its executives understand that noise will be one of the biggest barriers to expanding operations. Engineering teams around the world are at work redesigning drone propellers for a slower rotation in order to soften sound, or shrouding rotors with sound-absorbing nanofiber materials.

Blomberg dismisses most of these noise-reduction efforts as Trojan horses that will lull us into accepting the flying machines. He thinks the sheer numbers of drones will soon overwhelm well-intentioned engineering and routing schemes meant to mitigate their assault on our ears. Instead, Blomberg favors drone-delivery bans by municipalities, moratoria on deploying them, or if all else fails, at least "Sabbath days" with no flights allowed, so people can remember the world as it existed without the high-pitched whines.

"If we let these things in, and let them loose everywhere," he said, "they will radically change our soundscape."

Blomberg's "just say no" stance on drones once again raises that fundamental question about noise and noise control: what is our goal? For many, the answer is simply a quieter world, one with fewer decibels of extraneous sound careening around and disturbing our peace.

Quiet is indeed vital, and it's disappearing. Sonic tranquility is fast becoming a luxury product for those who can afford it.

At the same time, however, acoustic engineering and other proactive noise mitigations are not simply a back door for the proliferation of noisy industries, products, and practices. Without such efforts, we'd all still drive our cars, mow our lawns, travel by air, use power tools, and enjoy live music and sporting events. We would all keep making noise because that's what people do— only we'd make more of it.

More fundamentally, we could ultimately bolster the anti-noise cause by expanding soundscape goals beyond quiet and recognizing other measures of sonic quality besides low decibels. When the anti-noise agenda is limited to saying no, stop it, and shut up, it becomes easier for people to dismiss concerns about noise as subjective grievances, rather than a larger, societal issue we should tackle collectively and proactively, with more consequential ends than alleviating individual annoyance and bother.

We've seen how the neglect of sound helped noise take over the world, and we need to recognize that sounds matter. To care about chronic stress, distraction, sleep deprivation, patient safety, environmental justice, and habitat loss is to care about noise. So let's weave sound into broader conversations about all the big issues it touches. Yes, we should do more to protect ourselves from dangerous loudness and continue to practice common-sense noise control, but let's marry those efforts with a more positive soundscape agenda for a shared world that is, after all, filled with and defined by sound.

Sometimes we may want the world to shut up. But we ultimately want sonic environments that are more responsive to our needs. We want health and wellness. We want to feel safe, curious, joyful, and grounded. At times, we want to focus and be alone with our thoughts, but at others to connect with one another, to hear and be heard. Even though reality will inevitably fall short of our ideals, we can enlist sound in our efforts to create healthier buildings, restorative cities, and more resilient and equitable communities.

What will these places look like? What will they sound like?

ACKNOWLEDGMENTS

Book writing often feels like a lonely pursuit, and so it's both heartening and humbling to consider how many people helped me in big and small ways to make this particular book a reality.

A special shoutout to Erica Walker, an epidemiologist at Brown University and leader of the Community Noise Lab. The feature I wrote for the *Boston Globe Magazine* about Erica's research and work on behalf of noise-rattled Bostonians introduced me to the vastness of this topic, and I continue to benefit from her counsel. My other early explorations of noise were sustained by assignments from *Popular Science, Undark,* and *High Country News* and made possible by grants from the Pulitzer Center on Crisis Reporting and the Solutions Journalism Network.

When I proposed this book, my agent, David Patterson, found the perfect home for it at Norton. Thank you to Quynh Do for taking a chance on the project, and a massive thank you to my editor, Jessica Yao, for helping me find the signal in the noise of my early drafts. My writing was also much improved by the copyediting of Susanna Brougham and the fact checking of Hilary McClellen. Thanks to Abe Sands for deftly handling the lone visual featured in a sound-obsessed book.

I am deeply indebted to the scientists, scholars, and other experts who taught me about sound and noise from the cochlea on out, including Joe Allen, Jamie Banks, Frank Block Jr., Christopher Bonafide, Jennifer Jo Brout, Joan Casey, Judy Edworthy, Marcia Jenneth Epstein, Daniel Fink, Leila Hatch, Mack Hagood, Laurie Hanin, Gordon Hempton, Valtteri Hongisto, Dan Gauger, Lisa Goodrich, Helena Jahncke, Simon Jennings, Jacob Job, Gerald Kidd, Jordan Lacey, Ellen Lafargue, Nick Lesica, Bernie Krause, Lauren Kuehne, Alistair MacDonald, Megan McKenna, Kathy Metcalf, Stephan Moore, Nancy Nadler, Michael Osborne, Susan Parks, Sarah Payne, Antonella Radicchi, Susan Rogers, Zachary Rosenthal, Yoko Sen, Gregory Scott, Alex Smalley, and David Sykes.

In-person reporting is the lifeblood of a book like this. But the project began at the dawn of the Covid pandemic, and so I owe a special thanks to everyone who agreed to meet with me as the public health situation evolved. I gained a sense of sound's fundamental power from Deanna Meinke at the University of Northern Colorado. The musician and activist Kathy Peck, once and forever punk, let me tag along in San Francisco as she advised budding musicians about why and how they should protect their hearing. I learned about the intricate connections between ears and brains by interviewing Daniel Polley and Charles Liberman at Mass. Eye and Ear.

After shutting me in a pitch-black anechoic chamber for what seemed like an eternity (but was really just a minute), Joe Schlesinger, an anesthesiologist at Vanderbilt University Medical Center, told me about the risks of noisy operating rooms and alarm fatigue and how his lab works to improve the soundscapes of healing. I discovered how noise can harm the developing mind through the research of Nina Kraus and her crew at Northwestern University's Brainvolts lab, as well as from Arline Bronzaft, "the noise queen of New York City," who walked me through her groundbreaking studies of noise disrupting classroom learning in schools.

Along with Bronzaft, Les Blomberg of the Noise Pollution Clearinghouse was an invaluable resource on the short-lived anti-noise pollution efforts by the US Environmental Protection Agency. Blomberg, who thinks we should do a lot more to anticipate the next noise battle, generously agreed to meet with me and the drone-noise expert Eddie Duncan in Burlington, Vermont, where we looked out over Lake Champlain and imagined a future sky swarming with the high-pitched buzz of airborne delivery vehicles.

Thanks to Sumaira Abdulali, founder of the Awaaz Foundation in Mumbai, and to Yeshwant Oke, a doctor and pioneer of India's anti-noise efforts, for giving me a glimpse of noise pollution's global scale. I also learned a lot about environmental noise hazards around the world from Rick Neitzel at University of Michigan's School of Public Health. Closer to home, Neitzel's colleague, Stuart Batterman, shared his research on the trucking noise battering the residents of Detroit, people like Thomasenia Weston, who sat with me on her front porch and shared her story between the roars of big rigs passing by.

Other Detroiters deserving gratitude include Chavron Lyon, Simone Sagovac, Kenneth McPhail, Rico Razo, Diane Cheklich, and Ventra Asana. In New York City, I was privileged to meet and chat with Tanya Bonner, who fights noise on behalf of her neighbors in Washington Heights as cofounder of the WaHi-Inwood Task Force on Noise; I'm just as grateful to the members of the WaHi and Inwood for Respectful Decibel Levels Facebook Group, who agreed to talk with me. Thanks also to Maria Batayola of El Centro de la Raza in Seattle's Beacon Hill neighborhood, who weaves noise pollution into a broader battle for environmental justice.

On the docks of Woods Hole on Cape Cod, Aran Mooney and his lab team were kind enough to let me pester them about the unique sensory worlds of aquatic creatures, including the squid they were prepping for an underwater noise study. During my visit to the rocky shores of Monterey Bay, Brandon Southall was equally gracious as he educated me about chronic noise from shipping and the habitat risks it poses. In Fort Collins, Colorado, at the National Park Service's Natural Sounds and Night Skies Division, Emma Brown let me eavesdrop on America's wildest places and explained the work she and her colleagues do to protect these soundscapes. A hike-and-talk with Brown's former colleague Kurt Fristrup enlightened me about how noise in nature harms both animal and human listeners.

I benefited immensely from the insights of sound designers such as Joel Beckerman at Made Music, who introduced me to the concepts of "sonic trash" and composing experiences; Dallas Taylor, whose podcast Twenty Thousand Hertz is a celebration of sound's possibilities; and Evan Benway, founder of Moodsonic, who took me behind the scenes of creating unique soundscapes for offices, health care, and other high-stress environments. Thanks to acousticians Ben Markham of Acentech and Raj Patel of Arup for letting me experience the 3D listening rooms that let clients design by listening to future buildings, and to the team at the Well Living Lab in Rochester, Minnesota, who dig into the health and well-being impacts of sound and other environmental variables in homes, schools, and offices.

I spent some quality time in Montreal with Catherine Guastavino and others at McGill University's Sounds in the City Lab, who showed off their digital design tools and field demonstrations aimed at making better-sounding cities. Across the pond, Lisa Lavia of the UK Noise Abatement Society gave me an extensive listening tour of Brighton, her adopted home, and explained her conversion into an ambassador for soundscapes. Lavia also introduced me to Grant Waters at Anderson Acoustics, and his expansive efforts to help create a new sound for London's expanding fleet of electric buses. I also learned a great deal from University College London researchers Francesco Aletta, Tin Oberman, and Andrew Mitchell, key members of the Soundscape Indices project led by Jian Kang. Another UCL researcher, Nick Tyler, kindly invited me to witness multisensory urban-design research on a grand scale at his Person Environment Activity Research Laboratory, which had only recently opened and still had that new lab smell.

My reporting relied on far too many books to acknowledge by name, but a few were particularly useful. Much of what I know about the earliest years of decibel-based noise control and its antecedents came from *The Soundscape of Modernity* by Emily Thompson. And whenever I had a nagging question about anything noise related, my go-to was Hillel Schwartz's encyclopedic *Making Noise: From Babel to the Big Bang and Beyond.*

The Tuning of the World by R. Murray Schafer, along with the collection of acoustic ecology documents archived online by the World Soundscape Project at Vancouver's Simon Fraser University, was my essential primer on soundscapes and Schafer's call to take sound seriously, and not just noise.

As I reported, wrote, and rewrote, my family and friends likely learned more than they cared to know about the many milestones on the road to publication. Their patience, love, and support mean more to me than I can say.

NOTES

Prologue

2 **damage from excessive noise:** "Noise-Induced Hearing Loss," Centers for Disease Control and Prevention, https://www.cdc.gov/ncbddd/hearingloss/noise.html.

2 **depression and dementia:** David J. Mener et al., "Hearing Loss and Depression in Older Adults," *Journal of the American Geriatric Society* 61 (September 2013): 1627–29. Alison R. Huang et al., "Hearing Loss and Dementia Prevalence in Older Adults in the US," *JAMA* 329, no. 2 (January 2023): 171–73.

2 **hinder classroom learning:** Nina Kraus, *Of Sound Mind* (Cambridge, MA: MIT Press, 2021), 205–23.

2 **peers in quieter classrooms:** Arline L. Bronzaft and Dennis P. McCarthy, "The Effect of Elevated Train Noise on Reading Ability," *Environment and Behavior* 7, no. 4 (December 1975): 517–28.

2 **scores had evened out:** Arline L. Bronzaft, "The Effect of a Noise Abatement Program on Reading Ability," *Journal of Environmental Psychology* 1, no. 3 (September 1981): 215–22.

3 **premature deaths:** Eulalia Peris et al., "Environmental Noise in Europe—2020," European Environment Agency, EEA Report, no. 22 (March 2020).

3 **more than 100 million:** American Public Health Association, "Noise as a Public Health Hazard," Policy #202115, October 26, 2021.

3 **top environmental threats:** World Health Organization, "Noise," https://www.who.int/europe/news-room/fact-sheets/item/noise.

3 **1.4 billion motor vehicles:** "Automobile," Britannica online, updated June 24, 2024, https://www.britannica.com/technology/automobile.

4 **the spread of noise:** Noise Pollution Clearing House, "Noise Pollution in the 21st Century," https://www.nonoise.org/noise21cen.htm.

4 **increasingly urbanized souls:** United Nations, "Peace, Dignity, and Equality on a Healthy Planet."

5 **one-trillionth the energy:** "Decibel," Britannica online, https://www.britannica.com/science/decibel.

8 **new "quiet areas":** "Berlin Wird Leiser," website of Antonella Radicchi, creator of Hush City app, http://www.antonellaradicchi.it/portfolio/berlin-wird-leiser.

8 **"Our lives are enhanced":** Hannah Blythyn, minister for environment, "Noise and Soundscape Action Plan," Welsh Ministry of the Environment, 2018.

Chapter 1. Huh?!: Hearing Connections

11 **as if casting a spell:** Deanna Meinke, interview by author, June 25, 2021. Unless otherwise noted, all quotations from Meinke in this chapter are drawn from this interview.

12 **producing a shock wave:** Aatish Bhatia, "The Loudest Sound Ever Heard," *Discover*, July 13, 2018.

12 **top 73 million:** Adele M. Goman, Nicholas S. Reed, and Frank R. Lin, "Addressing Estimated Hearing Loss in Adults in 2060," *JAMA Otolaryngology—Head and Neck Surgery* 143, no. 7 (July 2017): 733-34.

12 **about one in four:** World Health Organization, "Deafness and Hearing Loss," updated February 2, 2024, https://ww.who.int/news-room/fact-sheets/detail/deafness-and-hearing-loss.

13 **20 and 20,000 hertz:** Dale Purves et al., "The Audible Spectrum," in *Neuroscience*, 2nd ed. (Sunderland, MA: Sinauer Associates, 2001).

14 **flow of neural messages:** Daniel Polley, professor of otolaryngology–head and neck surgery, Harvard Medical School; director of Eaton-Peabody Laboratories; interview by author, February 5, 2020.

14 **the buzz of tinnitus:** National Institutes of Health, "Cochlear Nerve Damage Associated with Tinnitus," updated January 9, 2024, NIH Research Matters.

14 **limit of 85 decibels:** National Institute for Occupational Safety and Health (NIOSH), "Noise and Hearing Loss."

14 **a fifty-year-old study:** NIOSH, *Criteria for a Recommended Standard: Occupational Exposure to Noise*, January 1972.

15 **70 decibels (dishwasher loud):** Monica S. Hammer, Tracy K. Swinburn, and Richard L. Neitzel, "Environmental Noise Pollution in the United States: Developing an Effective Public Health Response," *Environmental Health Perspectives* 122, no. 2 (February 2014): 115-19.

15 **cut the exposure time:** NIOSH, "Noise and Hearing Loss."

15 **one hour at 85 decibels:** Centers for Disease Control and Prevention, "Loud Noise Can Cause Hearing Loss."

15 **occupational hearing loss:** Elizabeth A. Masterson et al., "Trends in Worker Hearing Loss by Industry Sector, 1981–2010," *American Journal of Industrial Medicine* 58, no. 4 (April 2015): 392–401.

15 **up to 111 decibels:** Richard L. Neitzel et al., "Noise Levels Associated with New York City's Mass Transit Systems," *American Journal of Public Health* 99, no. 8 (August 2009): 1393–99.

16 **a peak of 116 decibels:** Sumi Sinha et al., "Cycling Exercise Classes May Be Bad for Your (Hearing) Health," *Laryngoscope* 127, no. 8 (October 12, 2016): 1873–77.

16 **they averaged 112 decibels:** Sarah M. Tittman et al., "No Shortage of Decibels in Music City: Evaluation of Noise Exposure in Urban Music Venues," *Laryngoscope* 131, no. 1 (January 2021): 25–27.

16 **"emerging issue of environmental concern":** United Nations Environment Programme, Frontiers 2022: Noise, Blazes, and Mismatches—Emerging Issues of Environmental Concern, Nairobi, Kenya, 2022.

16 **noise-induced hearing loss:** Jiena Zhou et al., "Occupational Noise-Induced Hearing Loss in China: A Systematic Review and Meta-analysis," *BMJ Open Access* 10, no. 9 (September 28, 2020): e039576.

17 **study of "silence zones":** Nitin Garg, "Study on the Establishment of a Diversified National Ambient Noise Monitoring Network in Seven Major Cities of India," *Current Science* 113, no. 7 (October 10, 2017): 1367–83.

17 **the gold standard:** Donald A. Vogel, Patricia A. McCarthy, Gene W. Bratt, and Carmen Brewer, "The Clinical Audiogram: Its History and Current Use," *Communicative Disorders Review* 1, no. 2 (2007): 81–94.

18 **"An audiogram is a very":** M. Charles Liberman, interview by author, November 19, 2021. Unless otherwise noted, all quotations from Liberman in this chapter are drawn from this interview.

18 **severe hearing loss:** Frank R. Lin et al., "Hearing Loss and Incident Dementia," *Archives of Neurology* 68, no. 2 (February 2011): 214–20.

19 **raised the risk of depression:** Blake J. Lawrence et al., "Hearing Loss and Depression in Older Adults: A Systematic Review and Meta-analysis," *Gerontologist* 60, no. 3 (2020): e137–e154.

19 **unblemished audiograms:** George A. Gates et al., "Hearing in the Elderly: The Framingham Cohort, 1983–1985," *Ear and Hearing* 11, no. 4 (August 1990): 247–56. Douglas L. Beck, David R. Larsen, and Erin J. Bush, "Speech in Noise: Hearing Loss, Neurocognitive Disorders, Aging, Traumatic Brain Injury, and More," *Journal of Otolaryngology-ENT Research* 10, no. 4 (2018): 199–205.

19 **hours of high-intensity noise:** Sharon G. Kujawa and M. Charles Liberman, "Adding Insult to Injury: Cochlear Nerve Degeneration After 'Temporary' Noise-Induced Hearing Loss," *Journal of Neuroscience* 28, no. 45 (November 11, 2009): 14077–85.

20 **about two dozen people:** Kujawa and Liberman, "Adding Insult to Injury,"
    14077-85.

21 **multi-talker conversations:** E. Colin Cherry, "Some Experiments on the Rec-
    ognition of Speech, with One and with Two Ears," *Journal of the Acoustical
    Society of America* 25, no. 5 (September 1953): 975-79.

22 **"progenitor cells":** Edwin W. Rubel, Stephanie A. Furrer, and Jennifer S.
    Stone, "A Brief History of Hair Cell Regeneration Research and Speculations
    on the Future," *Hearing Research* 297 (2013): 42-51.

23 **"It's like having a big box":** Lisa Goodrich, interview by author, February 26,
    2021.

24 **over-the-counter hearing aids:** US Congress. "Text—H.R. 2430—115th Con-
    gress (2017-18): FDA Reauthorization Act of 2017," August 18, 2017.

24 **over-the-counter devices:** US Food and Drug Administration, "OTC Hearing
    Aids: What You Should Know," updated May 3, 2023, https://www.fda.gov/
    medical-devices/hearing-aids/otc-hearing-aids-what-you-should-know.

25 **which is noise:** Nicholas Lesica, professor of neuroengineering, The Ear Insti-
    tute, University College London; interview by author, May 16, 2022.

26 **noise-exposed workers:** Deirdre R. Green, Elizabeth A. Masterson, and
    Christa L. Themann, "Prevalence of Hearing Protection Device Non-Use
    Among Noise-Exposed Workers in 2007 and 2014," *American Journal of
    Industrial Medicine* 64, no. 12 (October 2021): 1002-17.

28 **until middle age or later:** The Melanoma Research Alliance, "Melanoma Sta-
    tistics," https://tinyurl.com/2vn6czpv.

28 **hearing aids in her twenties:** Kathy Peck, interview by author, March 4, 2020.

29 **protection available to patrons:** Elon Ullman, Kathy Peck, and Helen Simon,
    "'Plugs in the Clubs' Initiative," *Canadian Audiologist* 2, no. 3 (May 14, 2015): 1-5.

29 **passed a similar ordinance:** Elon Ullman et al., "'Plugs in the Clubs' Initia-
    tive," 1-5.

29 **protection at every show:** "10C X Musicares Hearing Health," PearlJam.Com,
    https://pearljam.com/news/10c-x-musicares-hearing-health.

29 **started selling hearing protection:** "House of Blues Promotes Hearing Health
    for Music Fans," *Hearing Review*, December 9, 2016.

29 **to protect their hearing:** Jieun Cha et al., "Increase in Use of Protective Earplugs
    by Rock and Roll Concert Attendees When Provided for Free at Concert Ven-
    ues," *International Journal of Audiology* 54, no. 12 (November 2015): 984-86.

29 **hearing protection at a venue:** Elizabeth Francis Beach, Lillian Nielsen, and
    Megan Gilliver, "Providing Earplugs to Young Adults at Risk Encourages Pro-
    tective Behaviour in Music Venues," *Global Health Promotion* 23, no. 2 (2016):
    45-56.

29 **earplug vending machines:** Mira Miller, "A Local Startup Is Putting Earplug
    Vending Machines Inside Toronto Clubs," *BlogTO*, posted February 2, 2020.
    https://www.blogto.com/music/2020/02/whut-earplug-vending-machines
    -toronto.

30 **"The Hottest Thing to Wear":** Aliza Abarbanel, "The Hottest Thing to Wear to the Club Is a Pair of Earplugs," *GQ*, September 30, 2021.

30 **"Jolene Cookbook":** Genevieve Y. Martin, et al., "Jolene Cookbook." Oregon Health & Science University, 2021. https://htf.crearecomputing.com/wp-content/uploads/2021/12/Jolene-Cookbook-V3.1-sm.pdf.

30 **Apple Watch series 5:** Richard L. Neitzel et al., "Toward a Better Understanding of Nonoccupational Sound Exposures and Associated Health Impacts: Methods of the Apple Hearing Study," *Journal of the Acoustical Society of America* 151, no. 3 (March 2022): 1476–89.

30 **they are adjustable:** Toolkit for Safe Listening Devices and Systems. Geneva: World Health Organization and International Telecommunication Union, 2019. License: CC BY-NC-SA 3.0 IGO.

31 **a few topline findings:** University of Michigan School of Public Health, "Interactive US Maps of Noise Exposure," updated April 26, 2023.

Chapter 2. Hearing Ourselves Think: Noise Distraction

34 **the word for "ship":** Hillel Schwartz, *Making Noise* (Brooklyn, NY: Zone Books, 2016), 652.

35 **Typists in the quieter:** Donald A. Laird, "Experiments on the Physiological Cost of Noise," *Journal of the National institute of Industrial Psychology* 4 (1929): 251–58.

35 **the "Tensionometer":** Jam Handy Organization, "Let Yourself Go," January 1, 1940, Detroit, Michigan, https://archive.org/details/LetYours1940.

35 **the conversations of others:** William J. Cavanaugh et al., "Speech Privacy in Buildings," *Journal of the Acoustical Society of America* 34 (1962): 475–92.

35 **"speech transmission index":** Tammo Houtgast and Herman Steeneken, "A Physical Method for Measuring Speech-Transmission Quality," *Journal of the Acoustical Society of America* 67 (1980): 318–26.

36 **the notorious halfalogue:** Lauren L. Emberson, Gary Lupyan, Michael H. Goldstein, and Michael J. Spivey, "Overheard Cell-Phone Conversations: When Less Speech Is More Distracting," *Psychological Science* 21, no. 10 (2010): 1383–88. John E. Marsh et al., "Why Are Background Telephone Conversations Distracting?" *Journal of Experimental Psychology: Applied* 24, no. 2 (2018): 222–35.

36 **enough to become unintelligible:** Marsh et al., "Telephone Conversations."

37 **"The attitude was":** Nate Berg, "These Architects Popularized the Open Office. Now They Say 'The Open Office Is Dead,'" *Fast Company*, April 19, 2021.

38 **at least 70 percent:** Sarah Mravec and Diane Stegmeier, "The State of the Open Office," *FMJ* (Facilities Management Journal), January–February 2017, 20–23.

38 **"Unsurprisingly, then, very few":** "When the Walls Come Down: How Smart Companies Are Rewriting the Rules of the Open Workplace," *Oxford Economics*, June 13, 2016.

38 **follow-up survey two years later:** "The Unmet Promises of the Open-Plan Office," *Oxford Economics*, June 12, 2018.

38 **lack of speech privacy:** Center for the Built Environment, "Acoustical Analysis in Office Environments Using POE Surveys."

39 **about 150 employees:** Ethan S. Bernstein and Stephen Turban, "The Impact of the 'Open' Workspace on Human Collaboration," *Philosophical Transactions B* 373: 20170239 (2018).

40 **the same square footage:** Jesus Diaz, "The Real Reason Your Company Switched to an Open Plan Office," *Fast Company*, August 17, 2018.

41 **bad for the bottom line:** Andrea Gerlitz and Marcel Hülsbeck, "The Productivity Tax of New Office Concepts: A Comparative Review of Open-Plan Offices, Activity-Based Working, and Single-Office Concepts," *Management Review Quarterly* (January 2023): 1–31.

41 **averaging only 55 decibels:** Gary W. Evans and Dana Johnson, "Stress and Open-Office Noise," *Journal of Applied Psychology* 85, no. 5 (2000): 779–83.

41 **took more breaks:** Annu Haapakangas, Riikka Helenius, Esko Keskinen, and Valtteri Hongisto, "Perceived Acoustic Environment, Work Performance, and Well-Being—Survey Results from Finnish Offices." In *Ninth International Congress on Noise as a Public Health Problem (ICBEN)*, 18, no. 8 (2008): 21–25.

42 **significantly more sick leave:** Daniel Mauss, Marc N. Jarczok, Bernd Genser, and Raphael Herr, "Association of Open-Plan Offices and Sick Leave—A Systematic Review and Meta-analysis," *Industrial Health* 61 (2023): 173–83. Morten Birkeland Nielsen and Stein Knardahl, "The Impact of Office Design on Medically Certified Sickness Absence," *Scandinavian Journal of Work and Environmental Health* 46, no. 3 (May 2020): 330–34. Loretta G. Platts, Aram Seddigh, Erik Berntson, and Hugo Westerlund, "Sickness Absence and Sickness Presence in Relation to Office Type: An Observational Study of Employer-Recorded and Self-Reported Data from Sweden," *PLoS ONE* 15, no. 4 (April 2020): 1–13.

42 **spent sick at home:** Thomas Clausen, Karl Bang Christensen, Thomas Lund, and Jesper Kristiansen, "Self-Reported Noise Exposure as a Risk Factor for Long-Term Sickness Absence," *Noise Health* 11, no. 43 (April–June 2009): 93–97. Jesper Kristiansen, "Is Noise Exposure in Non-Industrial Work Environments Associated with Increased Sickness Absence?" *Noise and Vibration Worldwide* 41, no. 5 (2010): 9–16.

42 **test scores go down:** Bridget M. Shield and Julie E. Dockrell, "The Effects of Environmental and Classroom Noise on the Academic Attainments of Primary School Children," *Journal of the Acoustical Society of America* 123, no. 1 (2008): 133–44. Maria Klatte, Kristin Bergstrom, and Thomas Lachmann, "Does Noise Affect Learning? A Short Review on Noise Effects on Cognitive Performance in Children," *Frontiers in Psychology* 4 (2013): 578.

42 **reading and math scores:** National Academies of Sciences, Engineering, and Medicine, "Assessing Aircraft Noise Conditions Affecting Student Learning," vol. 1, 2014.

42 **cognitive performance is hurt:** Maria Klatte, Markus Meis, Helga Sukowski, and August Schick, "Effects of Irrelevant Speech and Traffic Noise on Speech Perception and Cognitive Performance in Elementary School Children," *Noise and Health* 9, no. 36 (July–September 2007): 64–74. Glada Guerra et al., "Loudness and Intelligibility of Irrelevant Background Speech Differentially Hinder Children's Short Story Reading," *Mind, Brain, and Education* 15, no. 1 (October 2020): 77–87. Patrik Sörqvist, "Effects of Aircraft Noise and Speech on Prose Memory: What Role for Working Memory Capacity?" *Journal of Environmental Psychology* 30, no. 1 (2010): 112–18.

42 **undermine cognitive connections:** Nina Kraus, interview by author, October 25, 2021.

43 **"We need to be less cavalier":** Nina Kraus, *Of Sound Mind* (Cambridge, MA: MIT Press, 2021), 209.

43 **2021 review of research:** R.W.J. Mcleod, L. Myint-Wilks, S. E. Davies, and H. A. Elhassan, "The Impact of Noise in the Operating Theatre: A Review of the Evidence," *Annals of the Royal College of Surgeons of England* 103, no. 2 (2021): 83–87.

44 **their earliest investigations:** Ryan A. Stevenson et al., "Effects of Divided Attention and Operating Room Noise on Perception of Pulse Oximeter Pitch Changes: A Laboratory Study," *Anesthesiology* 118, no. 2 (2013): 376–81.

44 **"When that beep turns":** Joseph J. Schlesinger, interview by author, July 22, 2021.

44 **Schlesinger's initial follow-up:** Joseph J. Schlesinger, Ryan A. Stevenson, Matthew S. Shotwell, and Mark T. Wallace, "Improving Pulse Oximetry Pitch Perception with Multisensory Perceptual Training," *Society for Technology in Anesthesia* 118, no. 6 (June 2014): 1249–53.

45 **"These findings," they wrote:** Schlesinger et al., "Improving Oximetry Pitch Perception," 1249–53.

45 **The near miss occurred:** Alistair MacDonald, interview by author, August 27, 2021.

47 **ran a pilot study:** Alistair MacDonald and Joseph Schlesinger, "Canary in an Operating Room: Integrated Operating Room Music," *Proceedings of the Human Factors and Ergonomics Society Europe* (2018): 79–83.

47 **nearly 7 seconds faster:** Akash K. Gururaja et al., "Modulating Operating Room Music Volume with the CanaryBox: A Quality Improvement Initiative to Improve Anesthesia Clinicians' Response Times to Alarms to Improve Quality of Anesthetic Care," *Human Factors in Healthcare* 2 (November 2022): 100029.

47 **"That's huge," Schlesinger told:** Joseph J. Schlesinger, interview by author, May 11, 2023.

48 **a few hundred hertz:** Niklas Moeller, "Sound Masking 101: Understanding and Specifying Sound Masking Technology," *Architectural Record,* December 2015.

49 **strategies often backfired:** Valtteri Hongisto, director, Built Environment Research Group, Turku University of Applied Sciences, Turku, Finland, interview by author, June 17, 2021. Unless otherwise noted, all quotations from Hongisto in this chapter are drawn from this interview.

50 **someone working in silence:** Valterri Hongisto, "A Model Predicting the Effect of Speech of Varying Intelligibility on Work Performance," *Indoor Air* 15, no. 6 (December 2005): 458-68. Annu Haapakangas, Valterri Hongisto, and Andreas Liebl, "The Relation Between the Intelligibility of Irrelevant Speech and Cognitive Performance—A Revised Model Based on Laboratory Studies," *Indoor Air* 30, no. 6 (2020): 1130-1146.

50 **Hongisto achieved similar results:** Valterri Hongisto and Annu Haapakangas, "Effect of Sound Masking on Workers in an Open Office," *Proceedings of Acoustics* 8, no. 29 (2008): 537-42.

50 **get distraction distance accepted:** International Organization for Standardization, ISO 3382-3:2022, "Acoustics—Measurement of Room Acoustic Parameters—Part 3: Open Plan Offices," 2022.

50 **mix of anti-noise measures:** Mervi Ahola, Jorma Säteri, and Laura Sariola, "Revised Finnish Classification of Indoor Climate 2018," CLIMA 2019 Congress, *E3S Web of Conferences* 111 (May 2019).

51 **"In trying to fix":** Alistair MacDonald, "Healthcare's Broken Volume Knob," unpublished, shared with the author on August 27, 2021.

Chapter 3. Feel the Noise: How Sound Gets Under Your Skin

53 **"unwanted and/or harmful sound":** American Public Health Association, "Noise as a Public Health Hazard," Policy #202115, October 26, 2021.

55 **what humans can hear:** Juliette Volcler, *Extremely Loud: Sound as a Weapon,* trans. Carol Volk (New York: The New Press, 2013), 23-26.

55 **"very nearly lethal":** Volcler, *Extremely Loud,* 23-26.

56 **can make eyes twitch:** Catherine Rasgaitis, "Resonant Frequencies of the Human Body," *Medium,* September 20, 2022.

56 **Studies of intense infrasound:** Juliana Araújo Alves, Filipa Neto Paiva, Lígia Torres Silva, and Paula Remoaldo, "Low-Frequency Noise and Its Main Effects on Human Health—A Review of the Literature Between 2016 and 2019," *Applied Sciences* 10, no. 15 (2020): 5205. Geoff Leventhall, Peter Pelmear, and Stephen Benton, "A Review of Published Research on Low-Frequency Noise and Its Effects," UK Department for Environment, Food, and Rural Affairs (London: Defra Publications, 2023). Christos Baliatsas, Irene van Kamp, Ric van Poll, and Joris Yzermans, "Health Effects from Low-Frequency Noise and Infrasound in the General Population: Is It Time to Listen? A System-

atic Review of Observational Studies," *Science of the Total Environment* 557 (March 2016): 163–69.

55 **killer-whistle research:** Volcler, *Extremely Loud*, 26.

56 **"fell down dead":** Lyall Watson, *Supernature: The Natural History of the Supernatural* (London: Hodder and Stoughton, 1973), 93.

56 **guitarist Jimmy Page:** William S. Burroughs, "Led Zeppelin, Jimmy Page, and Rock Magic," *Crawdaddy*, June 1975. Retrieved from Jordan Potter, "When William S. Burroughs Interviewed Jimmy Page," *Far Out*, January 16, 2024, https://faroutmagazine.co.uk/william-s-burroughs-interviewed -jimmy-page.

57 **a 7-meter sewer pipe:** "Infrasonic—The Pipe," Sarah Angliss personal website, https://www.sarahangliss.com/infrasound-the-pipe.

58 **investigate infrasound toxicology:** National Institutes of Health, "Infrasound: Brief Review of the Toxicology Literature," November 2001.

58 **"Most studies reported":** National Institutes of Health, "Infrasound."

58 **evidence of success:** Jürgen Altmann, "Acoustic Weapons—A Prospective Assessment," *Science and Global Security* 9, no. 3 (2001): 165–234.

58 **"Although high-intensity infrasound":** James R. Jauchem and Michael C. Cook, "High-Intensity Acoustics for Military Nonlethal Applications: A Lack of Useful Systems," *Military Medicine* 172, no. 2 (2007): 182–89.

59 **break up kidney stones:** Krishna Ramaswamy et al., "Targeted Microbubbles: A Novel Application for the Treatment of Kidney Stones," *BJU International* 116, no. 1 (2015): 9–16.

59 **and blood clots:** Robert J. Siegel and Huai Luo, "Ultrasound Thrombolysis," *Ultrasonics* 48, no. 4 (August 2008): 312–20.

59 **"deterrent tone":** "LRAD: Maritime Safety," Genasys, https://genasys.com/ lrad-solutions/maritime-safety.

60 **toll on every bodily system:** "Chronic Stress Can Hurt Your Overall Health," ColumbiaDoctors, Columbia University Irving Medical Center, May 19, 2023.

61 **first observed clinically:** Andrea E. Cavanna and Stefano Seri, "Misophonia: Current Perspectives," *Neuropsychiatric Disease and Treatment* (August 2015): 2117–23.

61 **A 2017 brain-scan study:** Sukhbinder Kumar et al., "The Brain Basis for Misophonia," *Current Biology* 27, no. 4 (February 2017): 527–33.

62 **menu of coping strategies:** Jennifer Jo Brout, director of the International Misophonia Research Network, interview by author, April 30, 2021.

63 **world's leading killer:** Tawseef Dar et al., "Greater Neurobiological Resilience to Chronic Socioeconomic or Environmental Stressors Associated with Lower Risk for Cardiovascular Disease Events," *Circulation: Cardiovascular Imaging* 13, no. 8 (2020): e010337.

63 **risk for heart disease:** Wolfgang Babisch, "Updated Exposure-Response Relationship Between Road Traffic Noise and Coronary Heart Disease: A Meta-analysis," *Noise and Health* 16, no. 68 (2014): 1–9.

63 **other risk factors:** Lars Jarup et al., "Hypertension and Exposure to Noise Near Airports: The HYENA Study," *Environmental Health Perspectives* 116, no. 3 (2008): 329–33.

64 **the airport in Zurich:** Apolline Saucy et al., "Does Night-Time Aircraft Noise Trigger Mortality? A Case-Crossover Study on 24,886 Cardiovascular Deaths," *European Heart Journal* 42, no. 8 (2021): 835–43.

64 **10-decibel bump:** Federal Aviation Administration, "FAA History of Noise," updated March 29, 2022, https://www.faa.gov/regulations_policies/policy _guidance/noise/history.

64 **people felt "highly annoyed":** Federal Aviation Administration, "Neighborhood Environmental Survey," updated August 31, 2023, https://www.faa.gov/ regulations_policies/policy_guidance/noise/survey.

65 **complaints had multiplied:** "Noise Complaints by Town," Massport, https://www.massport.com/environment/noise-abatement/logan-airport/ complaints-by-towns.

65 **how much overflight noise:** FAA, "Neighborhood Environmental Survey."

65 **no policy changes:** Federal Aviation Administration, "Noise Policy Review," updated January 10, 2024, https://www.faa.gov/noisepolicyreview.

67 **makes us easy prey:** Mathias Basner, interview by author, April 23, 2020. Unless otherwise noted, all quotations from Basner in this chapter are drawn from this interview.

68 **according to the CDC:** Centers for Disease Control and Prevention, "About Sleep and Your Heart Health," https://www.cdc.gov/heart-disease/about/ sleep-and-heart-health.html

68 **$131 billion by 2032:** Precedence Research, "Sleep Aids Market Size 2022 to 2032."

68 **only 45 decibels:** Anke Marks, Barbara Griefahn, and Mathias Basner, "Event-Related Awakenings Caused by Noctorunal Transportation Noise," *Noise Control Engineering Journal* 56, no. 1 (2008): 52–62. Mathias Basner, Uwe Müller, and Eva-Maria Elmenhorst, "Single and Combined Effects of Air, Road, and Rail Traffic Noise on Sleep and Recuperation," *Sleep* 34, no. 1 (2011): 11–23.

68 **noise was piped in:** Michael G. Smith et al., "Traffic Noise-Induced Changes in Wake-Propensity Measured with the Odds-Ratio Product (ORP)," *Science of the Total Environment* 805, no. 20 (January 2022): 150191.

69 **high the next day:** Nour Makarem et al., "Effect of Sleep Disturbances on Blood Pressure," *Hypertension* 77, no. 4 (2021): 1036–46.

71 **protecting heart health:** American Heart Association, "Lifestyle Changes to Prevent a Heart Attack."

Chapter 4. The Noise Gap: Sound Pollution and Environmental Justice

73 **the fledgling Environmental Protection Agency:** Sidney A. Shapiro, "Lessons from a Public Policy Failure: EPA and Noise Abatement," *Ecology Law Quarterly* 19, no. 1 (1992): 1–61. Chuck Elkins, former EPA deputy assistant administrator for noise control programs, interview by author, April 21, 2020.

74 **commerce and convenience:** "Noise as a Public Health Hazard," APHA, 2021, https://apha.org/Policies-and-Advocacy/Public-Health-Policy-Statements/Policy-Database/2022/01/07/Noise-as-a-Public-Health-Hazard.

75 **global economy roars past:** Thomasenia Weston, interview by author, August 11, 2021. Unless otherwise noted, all quotations from Weston in this chapter are drawn from this interview.

75 **about 45,000 people:** Data Driven Detroit, "Southwest Detroit Neighborhood Profile." https://datadrivendetroit.org/files/SGN/SW_Detroit_Neighborhoods _Profile_2013_081913.pdf.

75 **national noise gap:** Joan A. Casey et al., "Race/Ethnicity, Socioeconomic Status, Residential Segregation, and Spatial Variation in Noise Exposure in the Contiguous United States," *Environmental Health Perspectives* 125, no. 7 (2017): 077017.

76 **A 2020 audit:** Timothy W. Collins, Shawna Nadybal, and Sara E. Grineski, "Sonic Injustice: Disparate Residential Exposures to Transport Noise from Road and Aviation Sources in the Continental United States," *Journal of Transport Geography*, 82 (2020): 102604.

76 **cities such as Minneapolis:** Tsegaye Habte Nega, Laura Chihara, Kimberly Smith, and Mallika Jayaraman, "Traffic Noise and Inequality in the Twin Cities, Minnesota." *Human and Ecological Risk Assessment: An International Journal* 19, no. 3 (2013): 601–19. Mona Ray, "Environmental Justice: Segregation, Noise Pollution, and Health Disparities near the Hartsfield-Jackson Airport Area in Atlanta." *Review of Black Political Economy* 50, no. 1 (2023): 18–34.

76 **cities in California:** Cesar O. Estien, Christine E. Wilkinson, Rachel Morello-Frosch, and Christopher J. Schell, "Historical Redlining Is Associated with Disparities in Environmental Quality Across California," *Environmental Science and Technology Letters* 11 (2024): 54–59.

76 **students who were nonwhite:** Timothy W. Collins, Sara E. Grineski, and Shawna Nadybal, "Social Disparities in Exposure to Noise at Public Schools in the Contiguous United States," *Environmental Research* 175 (2019): 257–65.

76 **daily border crossings:** Luiza C. Savage, "Land of the Freeloaders: The Battle for a New Cross-Border Bridge," *Maclean's*, May 21, 2015, https://macleans.ca/news/canada/land-of-the-freeloaders-the-battle-for-a-new-cross-border-bridge.

76 **truck noise in Southwest:** Stuart Batterman et al., "A Community Noise Survey in Southwest Detroit and the Value of Supplemental Metrics for Truck Noise," *Environmental Research* 197 (2021): 111064.

77 **the risk of investing:** "Mapping Inequality: Redlining in New Deal America," University of Richmond, https://dsl.richmond.edu/panorama/redlining.

77 **demolished two neighborhoods:** Nushrat Rahman, "Report: Detroiters Hurt by Black Bottom, Paradise Valley Decimation Deserve Reparations," *Detroit Free Press*, August 31, 2023.

78 **a lower HOLC grade:** Haley M. Lane, Rachel Morello-Frosch, Julian D. Marshall, and Joshua S. Apte, "Historical Redlining Is Associated with Present-Day Air Pollution Disparities in U.S. Cities," *Environmental Science and Technology Letters* 9 (2022): 345–50. Kaya Bramble et al., "Exposure Disparities by Income, Race, and Ethnicity, and Historic Redlining Grade in the Greater Seattle Area for Ultrafine Particles and Other Air Pollutants," *Environmental Health Perspectives* 131, no. 7 (July 2023): 077004. Issam Motairek, Zhuo Chen, Mohamed H.E. Makhlouf, Sanjay Rajagopalan, and Sadeer Al-Kindi, "Historical Neighborhood Redlining and Contemporary Environmental Racism," *Local Environment* 28, no. 4 (2023): 518–28. Lindsay Kephart, "How Racial Residential Segregation Structures Access and Exposure to Greenness and Green Space: A Review," *Environmental Justice* 15, no. 4 (2022): 204–13. Anthony Nardone, Kara E. Rudolph, Rachel Morello-Frosch, and Joan A. Casey, "Redlines and Greenspace: The Relationship Between Historical Redlining and 2010 Greenspace Across the United States," *Environmental Health Perspectives* 129, no. 1 (2021): 017006.

78 **"hazardous" road noise:** Abas Shkembi, Lauren M. Smith, and Richard L. Neitzel, "Linking Environmental Injustices in Detroit, MI, to Institutional Racial Segregation Through Historical Federal Redlining," *Journal of Exposure Science and Environmental Epidemiology* (2022): 1–10.

78 **diagnosed with asthma:** Prudence Kunyanga and Beth Anderson, "Detroit: The Current Status of Asthma Burden (2021 Update)," Michigan Department of Health and Human Services—Michigan Asthma Surveillance, Data and Reports, 2021.

79 **joined a nonprofit:** Southwest Detroit Community Benefits Coalition, "About Us," https://swdetroitcbc.org.

79 **within three decades:** Michigan Department of Transportation, "Detroit River International Crossing: Final Environmental Impact Statement," 2008, htttps://www.partnershipborderstudy.com/reports_us.html#feis.

80 **over the neighborhood:** Manola Secaira, "Can Beacon Hill Win the Fight for Quieter Skies and a Healthier Neighborhood?" *Crosscut*, June 10, 2019, updated July 5, 2019.

80 **a community action plan:** El Centro de la Raza, "Beacon Hill Air and Noise Pollution," https://www.elcentrodelaraza.org/air-noise-ej-hj. Maria Batayola, El Centro de la Raza environmental justice coordinator, interview by author, February 10, 2021.

80 **five Beacon Hill schools:** Beacon Hill Seattle Noise, "Noise Health Effects," https://beaconhillseattlenoise.org/noise-health-effects.

81 **heart trouble started:** Andrew W. Correia, Junenette L. Peters, Jonathan I. Levy, Steven Melly, and Francesca Dominici, "Residential Exposure to Aircraft Noise and Hospital Admissions for Cardiovascular Diseases: Multi-Airport Retrospective Study," *BMJ* 347 (2013).

81 **"People don't experience":** Maria Batayola, interview by author, February 10, 2021. Unless otherwise noted, all quotations from Batayola in this chapter are drawn from this interview.

81 **taking measurements of sound:** Roseanne M. Lorenzana et al., "Beacon Hill Seattle Noise Measurement Project," OSF, updated November 12, 2019. https://osf.io/fn5z6.

81 **median was 68 decibels:** Lorenzana et al., "Beacon Hill."

82 **polluted industrial estuary:** US Environmental Protection Agency, "Lower Duwamish Waterway Seattle, WA: Cleanup Activities," https://cumulis.epa.gov/supercpad/SiteProfiles/index.cfm?fuseaction=second.cleanup&id=1002020.

82 **higher rates of poverty and lower life expectancy:** Linn Gould and BJ Cummings, "Duwamish Valley Cumulative Health Impacts Analysis." Seattle, WA: Just Health Action and Duwamish River Cleanup Coalition/Technical Advisory Group (March 2013).

82 **twenty wooded acres:** King County Conservation Futures, "East Duwamish Greenbelt: Brick Pits," updated October 13, 2022, https://your.kingcounty.gov/dnrp/library/water-and-land/stewardship/conservation-futures/applications/Listof2022Awards.pdf.

82 **its anti-noise efforts:** Port of Seattle, "Airport Noise Programs," https://www.portseattle.org/environment/airport-noise-programs.

83 **"Fix the Harm":** Beacon Hill Council Seattle, "Fix the Harm Campaign," https://www.beaconhillcouncilseattle.org/fix-the-harm.

83 **wealthier and whiter:** Margaret M. Ramirez, "City as Borderland: Gentrification and the Policing of Black and Latinx Geographies in Oakland," *Environment and Planning D: Society and Space* 38, no. 1 (2020): 147-66. Brandon Harris, Alessandro Rigolon, and Mariela Fernandez, "'To Them, We're Just Kids from the Hood': Citizen-Based Policing of Youth of Color, 'White Space,' and Environmental Gentrification," *Cities* 107 (2020): 102885. Lam Thuy Vo, "They Played Dominoes Outside Their Apartment for Decades. Then the White People Moved in and Police Started Showing Up," *BuzzFeed News*, June 29, 2018.

83 **significantly higher rates:** Joscha Legewie and Merlin Schaeffer, "Contested Boundaries: Explaining Where Ethnoracial Diversity Provokes Neighborhood Conflict," *American Journal of Sociology* 122, no. 1 (2016): 125-61.

84 **played go-go music:** Marissa J. Lang, "The Music Will Go On: Go-Go Returns Days After a Complaint Silenced a D.C. Store," *Washington Post*, April 10, 2019.

84 **in DC in the 1970s:** Natalie Hopkinson, *Go-Go Live: The Musical Life and Death of a Chocolate City* (Durham, NC: Duke University Press, 2012).

84 **fight funding cuts:** Don't Mute DC, "Origins and Impact," https://www
.dontmutedc.com/origins-impact.

85 **"I get it. I can understand":** Tanya Bonner, interview by author, May 11, 2021.

85 **a 1972 sit-in:** El Centro de la Raza, "History and Evolution," https://www
.elcentrodelaraza.org/aboutus/history-evolution.

86 **first citywide noise report:** Erica Walker, "2016 Greater Boston Noise Report,"
Community Noise Lab, https://communitynoiselab.org/2016-noise-report-2.

87 **sewage overflow:** Richard M. Mizelle Jr., "A Slow-Moving Disaster—The Jackson Water Crisis and the Health Effects of Racism," *New England Journal of
Medicine* 388 (June 15, 2023): 2212-14.

87 **below the poverty line:** US Census Bureau, "Quick Facts: Jackson City, Mississippi," https://data.census.gov/profile/Jackson_city,_Mississippi?g=160XX00
US2836000.

87 **high-quality data on their exposures:** Community Noise Lab, "Community
Noise Lab in Mississippi," https://communitynoiselab.org/mississippi.

87 **"I don't like the idea":** Erica Walker, interview by author, June 9, 2022. Unless
otherwise noted, all quotations from Walker in this chapter are drawn from
this interview.

88 **carve out $48 million:** Aaron Mondry, "How Residents of Southwest Detroit
Fought For, and Won, a Historic Community Benefits Package," *Model D,* January 29, 2018.

88 **thousands of vacant houses:** City of Detroit, "Bridging Neighborhoods,"
https://detroitmi.gov/government/mayors-office/bridging-neighborhoods.

89 **alternative truck routes:** Bianca Garcia, "Held Hostage by Pollution: Residents of Southwest Detroit Hold Out Hope for Truck Route Ordinance Before
Councilwoman Leaves Office," *Planet Detroit,* September 23, 2021.

90 **"It's strictly a budgetary":** Rico Razo, interview by author, September 30, 2021.

91 **loud and vibrant Delray:** Paul Sewick, "The Origins and Demise of Delray,"
Curbed Detroit, May 3, 2018.

91 **buyouts and demolishing houses:** John Carlisle, "Is This the End of Delray?"
Detroit Free Press, December 7, 2017, updated December 11, 2017.

91 **environmental degradation:** Bridging Neighborhoods, "What Is My Delray
Score?" https://detroitmi.gov/sites/detroitmi.localhost/files/2018-12/Whats
-My-Delray-Score.pdf.

92 **idled most operations:** Kayla Clarke, "US Steel to Idle Zug Island Plant, Leading to 1,500 Layoffs," *Click On Detroit,* December 19, 2019.

92 **sulfur dioxide and nitrogen oxides:** Union of Concerned Scientists, "Ripe for
Retirement: The Case for Closing Michigan's Costliest Coal Plants," November
2012.

93 **shutting down in 2021:** Joshua Singer, "EPA Settlement with DTE Energy to
Reduce Air Pollution in Southeast Michigan," US Environmental Protection
Agency press release, May 14, 2020, updated June 22, 2023. "DTE Energy Retires
'Small but Mighty' River Rouge Power Plant," DTE press release, June 4, 2021.

Chapter 5. Sensory Smog: Nature Is Listening

95   **"sensory smog":** Elizabeth Preston, "A Growing Sensory Smog Threatens the Ability of Fish to Communicate, Navigate, and Survive," *Science*, June 20, 2019.

96   **"I've come to believe":** Kurt Fristrup, interview by author, June 24, 2021. Unless otherwise noted, all quotations from Fristrup in this chapter are drawn from this interview.

97   **a constant crackle:** Leila Hatch, "Oh Snap! What Tiny Shrimp Can Tell Us About Habitat Health," National Marine Sanctuaries, December 2021.

97   **thousands of kilometers:** Zoe Cormier, "The Loudest Voice in the Animal Kingdom," *BBC Earth*.

97   **partly by listening:** "Baby Corals Dance Their Way Home," *Research Station Carmabi News*, February 2010.

98   **make their own sounds:** "Discovery of Sound in the Sea," University of Rhode Island Graduate School of Oceanography, https://dosits.org/animals.

98   **more than 6 miles away:** Ian T. Jones et al., "Changes in Feeding Behavior of Longfin Squid (*Doryteuthis Pealeii*) During Laboratory Exposure to Pile Driving Noise," *Marine Environmental Research* 165 (March 2021): 105250.

99   **hear an approaching dolphin:** T. Aran Mooney, interview by author, September 28, 2021. Unless otherwise noted, all quotations from Mooney in this chapter are drawn from this interview.

99   **Squid could detect sound:** T. Aran Mooney et al., "Sound Detection by the Longfin Squid (*Loligo pealeii*) Studied with Auditory Evoked Potentials: Sensitivity to Low-Frequency Particle Motion and Not Pressure," *Journal of Experimental Biology* 213, no. 21 (2010): 3748–59.

100 **Squid fishing is big:** Andrew M. Scheld, "Economic Impacts Associated with the Commercial Fishery for Longfin Squid (*Doryteuthis pealeii*) in the Northeast U.S.," Science Center for Marine Fisheries, May 6, 2020.

100 **recordings of pile driving:** Jones et al., "Feeding Behavior of Longfin Squid," 102520.

102 **died from overheating:** Rick Weiss, "Whales' Deaths Linked to Navy's Sonar Tests," *Washington Post*, December 30, 2001.

102 **sonar's impact on wildlife:** Brandon Southall, interview by author, September 19, 2022. Unless otherwise noted, all quotations from Southall in this chapter are drawn from this interview.

103 **170 to 190 underwater decibels:** "Sources of Noise," *Science Notes*, produced by University of California, Santa Cruz, and California State University, Monterey Bay, https://sciencenotes.ucsc.edu/9601/OceanNoise/Noises.html.

103 **ambient sound levels:** George V. Frisk, "Noiseonomics: The Relationship Between Ambient Noise Levels in the Sea and Global Economic Trends," *Scientific Reports* 2, no. 1 (2012): 437.

103 **essentially, whale shouting:** Kate Madin and Cherie Winner, "Are Whales 'Shouting' to Be Heard?" *Oceanus: The Journal of Our Ocean Planet*, November 10, 2010.

103 **whales simply stopped:** Koki Tsujii et al., "Change in Singing Behavior of Humpback Whales Caused by Shipping Noise," *PLoS One* 13, no. 10 (October 2018): e0204112.

104 **researchers struggled to find:** Leila Hatch, marine ecologist, Stellwagen Bank National Marine Sanctuary; co-director, NOAA SanctSound Program; interview by author, July 23, 2021. Unless otherwise noted, all quotations from Hatch in this chapter are drawn from this interview.

104 **Numbers of right whales:** NOAA Fisheries, "North Atlantic Right Whale," updated March 1, 2024, https://www.fisheries.noaa.gov/species/north-atlantic-right-whale.

104 **having fewer calves:** NOAA, "North Atlantic Right Whale."

105 **billions of zooplankton:** Miriam Wasser, "As Right Whale Population Plummets, Focus Turns to Their Falling Birth Rates," *WBUR*, aired August 21, 2018.

105 **their acoustic range:** Leila Hatch et al., "Quantifying Loss of Acoustic Communication Space for Right Whales in and around a US National Marine Sanctuary," *Conservation Biology* 26, no. 6 (2012): 983–94.

105 **reclaim their sonic space:** Lauren Sommer, "Whales Get a Break as Pandemic Creates Quieter Oceans," *NPR*, aired July 20, 2020.

105 **pods of humpbacks:** Amorina Kingdon, "In the Absence of Cruise Ships, Humpbacks Have Different Things to Say," *Hakai*, September 2, 2021. Anthony Zurcher, "Why Whales in Alaska Have Been So Happy," *BBC*, August 4, 2021.

106 **ships in the Arctic:** Melanie Lancaster, "Pollution with an Easy Solution: Regulating Underwater Noise Pollution for a Healthy Arctic Ocean," *World Wildlife Fund*, February 19, 2021.

106 **raised the alarm:** Carlos M. Duarte et al., "The Soundscape of the Anthropocene Ocean," *Science* 371, no. 6529 (2021): eaba4658.

106 **since its founding in 1916:** Diane Liggett, "The Nature of Sound," National Park Service, US Department of the Interior, 1999, https://npshistory.com/publications/sound/nature-of-sound/the-nature-of-sound.pdf.

107 **air-tour management plans:** National Parks Service, "National Parks Air Tour Management Program," updated December 14, 2023, https://www.nps.gov/subjects/sound/air-tours-program.htm.

108 **nearly 47,000 hours:** Rachel T. Buxton et al., "Noise Pollution Is Pervasive in US Protected Areas," *Science* 356, no. 6337 (2017): 531–33.

109 **"phantom road":** Christopher J.W. McClure, "An Experimental Investigation into the Effects of Traffic Noise on Distributions of Birds: Avoiding the Phantom Road," *Proceedings of the Royal Society B* 280, no. 1773 (2013): 20132290.

109 **birds that had toughed it:** Heidi E. Ware, Christopher J.W. McClure, Jay D. Carlisle, and Jesse R. Barber, "A Phantom Road Experiment Reveals Traffic Noise Is an Invisible Source of Habitat Degradation," *Proceedings of the National Academy of Sciences* 112, no. 39 (2015): 12105-9.

110 **natural gas wells:** Jennifer N. Philips, Sarah E. Termondt, and Clinton D. Francis, "Long-Term Noise Pollution Affects Seedling Recruitment and Community Composition, with Negative Effects Persisting After Removal," *Proceedings of the Royal Society B* 288, no. 1948 (2021): 20202906.

112 **"I had some great conversations":** Emma Brown, interview by author, June 24, 2021.

112 **Quieter ships have been:** Russell Leaper, Martin Renilson, Veronica Frank, and Vassili Papastavrou, "Possible Steps Toward Reducing Impacts of Shipping Noise," *IWC-SC/61 E* 19 (2009): 2009.

112 **Research vessels designed:** Clear Seas Centre for Responsible Marine Shipping, "Underwater Noise and Marine Mammals," https://clearseas.org/underwater-noise.

113 **6-8 decibels quieter:** Martin Gassmann, Lee B. Kindberg, Sean M. Wiggins, and John A. Hildebrand, "Underwater Noise Comparison of Pre- and Post-Retrofitted MAERSK G-Class Container Vessels," Marine Physical Laboratory, Scripps Institution of Oceanography, University of California, San Diego; La Jolla, California, MPL TM-616, October 2017.

113 **IMO finally issued:** International Maritime Organization, "Ship Noise," https://www.imo.org/en/MediaCentre/HotTopics/Pages/Noise.aspx.

113 **"Unless everyone's doing":** Kathy Metcalf, interview by author, June 3, 2021.

113 **the IMO updated:** International Maritime Organization, "Addressing Underwater Noise from Ships—Draft Revised Guidelines Agreed," January 30, 2023, https://www.imo.org/en/MediaCentre/Pages/WhatsNew-1818.aspx.

114 **"slow steaming":** Container xChange, "How Slow Steaming Impacts Shippers and Carriers," December 16, 2019.

114 **endangered orcas and whales:** Vancouver Fraser Port Authority (VFPA), ECHO Program Summary Report 2021: Voluntary Vessel Slowdown in Haro Strait and Boundary Pass, June 2022.

115 **billions of dollars:** National Park Service, "Visitor Spending Effects," updated August 21, 2023, https://www.nps.gov/subjects/socialscience/vse.htm.

115 **listening to nature:** Amanda Rapoza, Erika Sudderth, and Kristin Lewis, "The Relationship Between Aircraft Noise Exposure and Day-Use Visitor Survey Responses in Backcountry Areas of National Parks," *Journal of the Acoustical Society of America* 138, no. 4 (2015): 2090-105.

115 **one-third less scenic:** David Weinzimmer et al., "Human Responses to Simulated Motorized Noise in National Parks," *Leisure Sciences* 36, no. 3 (2014): 251-67.

115 **minutes of recorded birdsong:** Emil Stobbe, Josefine Sundermann, Leonie Ascone, and Simone Kühn, "Birdsongs Alleviate Anxiety and Paranoia in Healthy Participants," *Scientific Reports* 12, no. 1 (2022): 16414.

115 **mood boosts in response:** Ryan Hammoud et al., "Smartphone-Based Ecological Momentary Assessment Reveals Mental Health Benefits of Birdlife," *Scientific Reports* 12, no. 1 (2022): 17589.

116 **health and happiness:** Rachel T. Buxton et al., "A Synthesis of Health Benefits of Natural Sounds and Their Distribution in National Parks," *Proceedings of the National Academy of Sciences* 118, no. 14 (2021): e2013097118.

117 **pile into the woods:** Lary M. Dilsaver, "Preservation Choices at Muir Woods," *Geographical Review* (1994): 290–305.

117 **amounts of human noise:** David W. Stack, Newman Peter, Robert E. Manning, and Kurt M. Fristrup, "Reducing Visitor Noise Levels at Muir Woods National Monument Using Experimental Management," *Journal of the Acoustical Society of America* 129, no. 3 (2011): 1375–80.

118 **visitors and birds alike:** Mitchell J. Levenhagen et al., "Ecosystem Services Enhanced Through Soundscape Management Link People and Wildlife," *People Nat* 3, no. 1 (2020): 176–89.

Chapter 6. Beyond Noise: A World of Unbounded Sound

123 **defeat the urban din:** Emily Thompson, *The Soundscape of Modernity: Architectural Acoustics and the Culture of Listening in America, 1900–1933* (Cambridge, MA: MIT Press, 2004), 157–68.

123 **nighttime chariot driving:** Edgar A.G. Shaw, "Noise Environments Outdoors and the Effects of Community Noise Exposure," *Noise Control Engineering Journal* 44, no. 3 (1996): 109–19.

123 **Victorian England:** John M. Picker, *Victorian Soundscapes* (New York: Oxford University Press, 2003).

124 **Society for the Suppression:** George Prochnik, *In Pursuit of Silence: Listening for Meaning in a World of Noise* (New York: Anchor Books, 2010), 204–16.

124 **first noise code in 1936:** NYC Department of Environmental Protection, "Applying the NYC Noise Code," https://www.nyc.gov/assets/dep/downloads/pdf/environment/education/noise-sound-applying-nyc-noise-code.pdf.

125 **R. Murray Schafer:** William Robin, "R. Murray Schafer, Composer Who Heard Nature's Music, Dies at 88," *New York Times*, August 23, 2021, updated August 28, 2021.

126 **background "keynote sounds":** R. Murray Schafer, "The Vancouver Soundscape," World Soundscape Project, Document No. 5 (Vancouver: Sonic Research Studio, Communications Studies Department, Simon Fraser University, 1973).

126 **"We must seek a way":** R. Murray Schafer, *The Tuning of the World: Toward a Theory of Soundscape Design* (New York: Knopf, 1977), 4.

127 **"The final question":** Schafer, *The Tuning of the World*, 5.

129 **more than tripling:** International Energy Agency, "Electric Vehicles," https://www.iea.org/energy-system/transport/electric-vehicles.

129 **cars sold by 2030:** Neil Winton, "European EV Sales Growth Slows, but 2030 Forecasts Remain Ambitious," *Forbes*, November 2, 2023. Nick Carey, "As Prices Fall, Two Thirds of Global Car Sales Could Be EVs by 2030, Study Says," *Reuters*, September 14, 2023.

129 **transport authority of London:** Anderson Acoustics, "TfL AVAS Urban Bus Sound," https://andersonacoustics.co.uk/case-study/soundscapes/tfl-avas -urban-bus-sound.

130 **"If you simply did":** Grant Waters, interview by author, July 18, 2022. Unless otherwise noted, all quotations from Waters in this chapter are drawn from this interview.

132 **from perplexed to disdainful:** Matthew Weaver, "Futuristic Sounds to Make Electric Buses Safer Hit Wrong Note," *Guardian*, July 1, 2019.

133 **psychoacoustic studies:** Berhard Seeber, "Psychoacoustics—Sound Quality: Sharpness, Fluctuation Strength, Roughness," Technical University of Munich, MOOC, "Fundamentals of Communication Acoustics," YouTube video, 8:55, January 28, 2020.

134 **fleet of electric buses:** Transport for London, "Bus Fleet Audit: 31 March 2023," https://content.tfl.gov.uk/fleet-annual-audit-report-31-march-2023.pdf.

135 **"People can quickly":** Joel Beckerman, interview by author, June 16, 2021. Unless otherwise noted, all quotations from Beckerman in this chapter are drawn from this interview.

135 **"sonic trash":** Joel Beckerman, "Scoring the World: A Systems-Thinking Approach to Sonic Branding and Design," *Music and the Moving Image* 13, no. 1 (2020): 3–20.

136 **the implicit association test:** Harvard University, "Project Implicit," https:// implicit.harvard.edu/implicit/takeatest.html.

137 **a 2018 Sentient study:** Kevin Perlmutter and Anjali Nair, "Cracking the Code on Sound in Experience Design," Made Music, 2018, https://mademusicstudio .com/sonicpulse-whitepaper.

Chapter 7. All the Machines That All Go Beep: Solving Signal Overload

139 **"a curious kind of nostalgia":** Emily Thompson, "The Roaring Twenties" (draft version), http://nycitynoise.com.

139 **"a complicated longing":** Thompson, "The Roaring Twenties."

140 **"peak beep":** Beth Teitell, "Second-Hand Beep Rage? It's a Thing," *Boston Globe*, April 29, 2019.

141 **"ubiquitous computing":** Mark Weiser, "The Computer for the 21st Century," *Scientific American*, September 1991, 94–104.

141 **"calm technology":** Peter J. Denning, Robert M. Metcalfe, Mark Weiser, and John Seely Brown, "The Coming Age of Calm Technology," *Beyond Calculation: The Next Fifty Years of Computing* (1997): 75–85.

141 **"If computers are everywhere":** Denning et al., "The Coming Age of Calm Technology," 75-85.

142 **risks of "alarm fatigue":** Christopher P. Bonafide et al., "Association Between Exposure to Nonactionable Physiologic Monitor Alarms and Response Time in a Children's Hospital," *Journal of Hospital Medicine* 10, no. 6 (2015): 345-51.

143 **alarm-related deaths:** Maria Cvach, "Monitor Alarm Fatigue: An Integrative Review," *Biomedical Instrumentation and Technology* 46, no. 4 (2012): 268-77.

143 **errors related to alarms:** Sasikala Thangavelu, Jasmy Yunus, Emmanuel Ifeachor, and Judy Edworthy, "Improving Alarm Response in ICU/CCU," *International Journal of Simulation—Systems, Science and Technology* 16, no. 4 (2015): 1.

143 **a danger to patients:** Sue Sendelbach and Marjorie Funk, "Alarm Fatigue: A Patient Safety Concern," *AACN Advanced Critical Care* 24, no. 4 (2013): 378-86.

143 **Patient Safety Learning Lab:** Christopher Bonafide, attending physician, Children's Hospital of Philadelphia; principal investigator, Patient Safety Learning Lab; interview by author, October 21, 2021.

143 **cut nuisance alarms:** Children's Hospital Association, "Patient Safety Project Aims to Reduce Alarm Fatigue in the NICU," April 2019, https://www .childrenshospitals.org/news/childrens-hospitals-today/2019/04/patient -safety-project-aims-to-reduce-alarm-fatigue-in-the-nicu.

143 **a follow-up article:** Kimberly Albanowski et al., "Ten Years Later, Alarm Fatigue Is Still a Safety Concern," *AACN Advanced Critical Care* 34, no. 3 (2023): 189-97.

144 **"The thinking is, well":** Joseph J. Schlesinger, interview by author, July 22, 2021. Unless otherwise noted, all quotations from Schlesinger in this chapter are drawn from this interview.

146 **more than 11 decibels:** Joseph J. Schlesinger et al., "Acoustic Features of Auditory Medical Alarms—An Experimental Study of Alarm Volume," *Journal of the Acoustical Society of America* 143, no. 6 (2018): 3688-97.

146 **an in-ear device:** "Device Helps ICU Patients by Filtering Out Noise from Medical Alarms," *VUMC Reporter*, June 21, 2017.

147 **the vibrational alerts:** May Gellert et al., "Comparing Auditory and Tactile Cues to Inform Clinicians of Patients' Vital Signs," in *Proceedings of the International Symposium on Human Factors and Ergonomics in Health Care*, vol. 9, no. 1, 193-94; Sage and Los Angeles, CA: Sage Publications, 2020. Parisa Alirezaee, Antoine Weill-Duflos, Joseph J. Schlesinger, and Jeremy R. Cooperstock, "Exploring the Effectiveness of Haptic Alarm Displays for Critical Care Environments," in *2020 IEEE Haptics Symposium (HAPTICS)*, 948-54, IEEE, 2020.

147 **last sense to go:** Elizabeth G. Blundon et al., "Electrophysiological Evidence of Preserved Hearing at the End of Life," *Scientific Reports* 10, no. 1 (2020): 10336.

148 **"Oh, don't worry":** Yoko Sen, interview by author, June 13, 2022. Unless otherwise noted, all quotations from Sen in this chapter are drawn from this interview.

148 **a cassette tape:** Sen Sound, "Beepers and the Beeped," Vimeo video of oral presentation at e-Forum Acusticum 2020, 12:54, December 2020.

149 **"I wish it to be":** Sen Sound, "Beepers and the Beeped."

149 **related to "functionality":** Avery Sen et al., "Functional and Sensible: Patient Monitoring Alarm Tones Designed with Those Who Hear Them," *Design Research Society*, June 25, 2022.

149 **"warned without being jolted":** Sen Sound, "Beepers and the Beeped."

150 **functionality and sensibility:** Sen et al., "Functional and Sensible."

150 **adopted the redesigned sounds:** Philips, "Philips Receives Red Dot: Best-of-the-Best Design Awards for Design Excellence," November 9, 2023, https://www.usa.philips.com/a-w/about/news/archive/standard/news/articles/2023/20231109-philips-receives-red-dot-best-of-the-best-design-awards-for-design-excellence.html.

150 **"the six ways people die":** Eric Boodman, "Anatomy of a Beep: A Medical Device Giant and an Avant-Garde Musician Set Out to Redesign a Heart Monitor's Chirps," *STAT*, September 10, 2018.

151 **well-known songs:** Frank Block, interview by author, April 27, 2021. Unless otherwise noted, all quotations from Block in this chapter are drawn from this interview.

151 **had previously designed alarms:** Boodman, "Anatomy of a Beep."

153 **nobody tested these:** Boodman, "Anatomy of a Beep."

153 **alerts easier to learn:** Judy Edworthy, interview by author, April 6, 2021. Unless otherwise noted, all quotations from Edworthy in this chapter are drawn from this interview.

154 **connections to the six ways:** Boodman, "Anatomy of a Beep."

154 **bested the beeps:** Judy Reed Edworthy, Cassie J. Parker, and Emily V. Martin, "Discriminating Between Simultaneous Audible Alarms Is Easier with Auditory Icons," *Applied Ergonomics* 99 (2022): 103609. Richard R. McNeer et al., "Auditory Icon Alarms Are More Accurately and Quickly Identified Than Current Standard Melodic Alarms in a Simulated Clinical Setting," *Anesthesiology* 129, no. 1 (2018): 58–66.

154 **The icons also prompted:** McNeer et al., "Auditory Icon Alarms," 58–66.

155 **patient-satisfaction surveys:** Laura Landro, "Hospitals Work on Patients' Most-Frequent Complaint: Noise," *Wall Street Journal*, June 10, 2013.

155 **Medicare reimbursement dollars:** Natalie Vaughn and Elizabeth Snively, "Maximizing Healthcare Reimbursement Through Higher Patient Satisfaction Scores," *Relias*, January 8, 2024.

155 **restful interludes:** Heidi Boehm and Stacy Morast, "Quiet Time: A Daily Period Without Distraction Benefits Both Patients and Nurses," *American Journal of Nursing* 109 (November 2009): 29-32.

156 **voices and laughter:** Stanford University, "Yoko Sen: What's the Last Sound You Want to Hear Before You Die?" YouTube video, 1:20, September 20, 2016.

Chapter 8. Sound Castles: Better Living Through Listening

160 **thousands of Manhattan restaurants:** Gregory Scott, "An Exploratory Survey of Sound Levels in New York City Restaurants and Bars," *Open Journal of Social Sciences* 6, no. 8 (2018): 64.

161 **complaints about restaurants:** "Zagat Releases 2018 Dining Trends Survey," Zagat Blog, January 8, 2018.

161 **unbalancing of restaurant acoustics:** Julia Belluz, "Why Restaurants Became So Loud—And How to Fight Back," *Vox*, July 27, 2018.

161 **"front-of-the-house culture":** Adam Platt, "Great Noise Boom," *Grub Street*, the food and restaurant blog of *New York Magazine*, July 13, 2013.

162 **just enough acoustic energy:** Sound-meter decibel measurements made by author on Sunday, November 20, 2022.

162 **17,000 square feet of sound absorption:** Benjamin Markham, president of Acentech, interview by author, July 9, 2021. Unless otherwise noted, all quotations from Markham in this chapter are drawn from this interview.

163 **computationally intense:** Raj Patel, principal of Arup; interview by author, June 13, 2022.

166 **Healthy Buildings program:** Healthy Buildings, "9 Foundations: Clear and Actionable Core Elements of Healthy Indoor Environments," Harvard T. H. Chan School of Public Health, https://9foundations.forhealth.org.

166 **Allen worked for his dad:** Joseph G. Allen, interview by author, April 1, 2021. Unless otherwise noted, all quotations from Allen in this chapter are drawn from this interview.

166 **"The 9 Foundations":** Joseph G. Allen et al., "The 9 Foundations of a Healthy Building," *Harvard: School of Public Health*, 2017.

167 **The noise pages:** Allen et al., "The 9 Foundations of a Healthy Building."

167 **"healthy building" certifications:** Natasha Sadikin, Irmak Turan, and Andrea Chegut, "The Financial Impact of Healthy Buildings: Rental Prices and Market Dynamics in Commercial Office," *MIT Center for Real Estate Research Paper* 21/04 (2021).

168 **in the visual sense:** Allen et al., "The 9 Foundations of a Healthy Building."

168 **nature brought indoors:** Jie Yin et al., "Effects of Biophilic Indoor Environment on Stress and Anxiety Recovery: A Between-Subjects Experiment in Virtual Reality," *Environment International* 136 (2020): 105427.

169 **silent forest was no better:** Matilda Annerstedt et al., "Inducing Physiological Stress Recovery with Sounds of Nature in a Virtual Reality Forest—Results from a Pilot Study," *Physiology and Behavior* 118 (2013): 240-50.

169 **three biophilic conditions:** Sara Aristizabal et al., "Biophilic Office Design: Exploring the Impact of a Multisensory Approach on Human Well-Being," *Journal of Environmental Psychology* 77 (2021): 101682.

173 **record-breaking heat wave:** Jacob Knutson, "London Declares 'Major Incident' After Fires Break Out During Record Heat," *Axios*, July 19, 2022.

173 **anonymous responses:** Stephen Hill, Vaimo's head of people and culture, interview by author, July 18, 2022. Unless otherwise noted, all quotations from Hill in this chapter are drawn from this interview.

174 **nature-adjacent offerings:** Jane Margolies, "Offices Dangle Beehives and Garden Plots to Coax Workers Back: The Latest Perks Include Harvesting Honey and Digging in the Dirt, Part of a Growing Effort to Give Employees Access to Fresh Air, Sunlight, and Plants," *New York Times*, August 25, 2021.

Chapter 9. Beyond Quiet: Hearing the Future City

179 **designate "quiet areas":** Chiara Bartalucci et al., "Quiet Areas Definition and Management in Action Plans: General Overview," in *INTER-NOISE and NOISE-CON Congress and Conference Proceedings*, 2012, no. 4: 7012-18. Institute of Noise Control Engineering, 2012.

179 **The banks of the Shannon:** Simon Jennings, interview by author, September 28, 2022. Unless otherwise noted, all quotations from Jennings in this chapter are drawn from this interview.

180 **set out to redefine quiet:** Hush City App, Antonella Radicchi personal website, http://www.antonellaradicchi.it/portfolio/hush-city-app.

180 **"everyday quiet areas":** Antonella Radicchi, "'Everyday Quiet Areas': What They Mean and How They Can be Integrated in Noise Action Plans," in *INTER-NOISE and NOISE-CON Congress and Conference Proceedings*, 258, no. 4, 3727-35. Institute of Noise Control Engineering, 2018.

181 **two distinct sound installations:** Daniel Steele et al., "Sounds in the City: Improving the Soundscape of a Public Square Through Sound Art," in *Proceedings of the ISCV*, 26, no. 8, 2019.

181 **The McGill team:** Valérian Fraisse, Daniel Steele, Simone d'Ambrosio, and Catherine Guastavino, "Shaping Urban Soundscapes Through Sound Art: A Case Study in a Public Square Exposed to Construction Noise," in *International Workshop on Haptic and Audio Interaction Design*, 2020.

181 **people rated traffic noise:** Joo Young Hong et al., "Effects of Adding Natural Sounds to Urban Noises on the Perceived Loudness of Noise and Soundscape Quality," *Science of the Total Environment* 711 (2020): 134571.

181 **birdsong or fountain sounds:** Joo Young Hong et al., "A Mixed-Reality Approach to Soundscape Assessment of Outdoor Urban Environments Augmented with Natural Sounds," *Building and Environment* 194 (2021): 107688.

182 **the Hush City app:** Antonella Radicchi, "Citizen Science Mobile Apps for Soundscape Research and Public Spaces Studies: Lessons from the Hush City Project," in Artemis Skarlatidou and Muki Haklay, eds., *Geographic Citizen Science Design: No One Left Behind.* UCL Press (2021): 130–48.

182 **"I wanted people to share":** Antonella Radicchi, interview by author, March 26, 2021. Unless otherwise noted, all quotations from Radicchi in this chapter are drawn from this interview.

182 **looking for nearby quiet:** Antonella Radicchi, "Hush City: A New Mobile Application to Crowdsource and Assess Everyday Quiet Areas in Cities," in *Proceedings of Invisible Places: The International Conference on Sound, Urbanism, and the Sense of Place*, 7–9, 2017, https://opensourcesoundscapes .org/wp-content/uploads/2018/04/Radicchi_2017_Hush-City-app.pdf.

182 **"everyday quiet areas" in Reuterkiez:** Antonella Radicchi, Dietrich Henckel, and Martin Memmel, "Citizens as Smart, Active Sensors for a Quiet and Just City. The Case of the 'Open-Source Soundscapes' Approach to Identify, Assess, and Plan 'Everyday Quiet Areas' in Cities," *Noise Mapping* 5, no. 1 (2016): 1–20. Antonella Radicchi, "Beyond the Noise: Open Source Soundscapes—A Mixed Methodology to Analyse, Evaluate, and Plan 'Everyday' Quiet Areas," in *Proceedings of Meetings on Acoustics*, 30, no. 1, AIP Publishing, 2017.

184 **The result was Annex 10:** Berlin Senatsverwaltung für Umwelt, Verkehr, und Klimaschutz. Lärmaktionsplan Berlin, 2019–23 Anlage 10: Ruhige Gebiete und städtische Ruhe- und Erholungsräume, Berlin 2020, https://www.berlin.de/ sen/uvk/umwelt/laerm/laermminderungsplanung-berlin/laermaktionsplan -2019-2023/download.

184 **"Small-scale retreats":** Berlin Senatsverwaltung für Umwelt, Verkehr, und Klimaschutz. Lärmaktionsplan Berlin, 2019–23 (translation by Google Translate).

185 **designating them as quiet:** Simon Jennings, email to author, February 21, 2024.

185 **a soundscape evangelist:** Lisa Lavia, interview by author, March 15, 2021.

186 **Brighton's city council:** Lisa Lavia et al., "Sounding Brighton: Practical Approaches Towards Better Soundscapes," in *INTER-NOISE and NOISE-CON Congress and Conference Proceedings*, 436–44. Institute of Noise Control Engineering, 2012.

186 **played at half speed:** Ben Beaumont-Thomas, "The Human League's Martyn Ware Has Used 3D Sound to Calm Brighton Drunks. Now He's Using It to Explore Alzheimer's," *Guardian*, February 4, 2014.

187 **"One principle of soundscape":** Lisa Lavia, interview by author, July 21, 2022. Unless otherwise noted, all further quotations from Lavia in this chapter are drawn from this interview.

189   **the entire universe:** André Fiebig, "Soundscape: A Construct of Human Perception," in *Soundscapes: Humans and Their Acoustic Environment*, pp. 23–48. Cham: Springer International Publishing, 2023.

189   **Swedish environmental psychologists:** Östen Axelsson, Mats E. Nilsson, and Birgitta Berglund, "A Principal Components Model of Soundscape Perception," *Journal of the Acoustical Society of America* 128, no. 5 (2010): 2836–46.

191   **an extensive menu:** Richard Yanaky, Darcy Tyler, and Catherine Guastavino, "City Ditty: An Immersive Soundscape Sketchpad for Professionals of the Built Environment," *Applied Sciences* 13, no. 3 (2023): 1611.

192   **"We might ask them":** Richard Yanaky, interview by author, September 15, 2021. Unless otherwise noted, all quotations from Yanaky in this chapter are drawn from this interview.

193   **funding from the European Union:** University College London, "Soundscape Indices (SSID)," August 24, 2018.

193   **perception for every individual:** Andrew James Mitchell et al., "The Soundscape Indices (SSID) Protocol: A Method for Urban Soundscape Surveys—Questionnaires with Acoustical and Contextual Information," *Applied Sciences* 10, no. 7 (2020): 2397.

193   **bearded, bespectacled doctoral student:** Andrew James Mitchell, interview by author, July 19, 2021. Unless otherwise noted, all quotations from Mitchell in this chapter are drawn from this interview.

195   **chunks of acoustic data:** Andrew James Mitchell, "Predictive Modeling of Complex Urban Soundscapes: Enabling an Engineering Approach to Soundscape Design," PhD diss., University College London, 2022.

195   **researchers still managed:** Andrew James Mitchell et al., "Investigating Urban Soundscapes of the COVID-19 Lockdown: A Predictive Soundscape Modeling Approach," *Journal of the Acoustical Society of America* 150, no. 6 (2021): 4474–88.

197   **dancing grannies:** Lily Kuo, "The Jig Is Up for China's Dancing Grannies Under New Noise Pollution Law," *Washington Post*, December 17, 2021.

198   **"house the world":** Nile Bridgeman, "Penoyre and Prasad's PEARL Research Facility," *Architects' Journal*, June 1, 2022.

198   **The main experimental space:** "PEARL," University College London, https://www.ucl.ac.uk/person-environment-activity-research-laboratory/welcome-pearl.

199   **"Tomorrow we're going":** Nick Tyler, interview by author, July 19, 2022. Unless otherwise noted, all quotations from Tyler in this chapter are drawn from this interview.

200   **enhance pedestrian safety:** Conrad Quilty-Harper, "The Radical Changes Coming to the City of London," *Bloomberg*, January 22, 2024.

200   **scooter collisions:** Transport for London, "Casualties in Greater London During 2022," September 2023, https://content.tfl.gov.uk/casualties-in-greater-london-2022.pdf.

201 **a background buzz:** City of London Corporation, Department of Markets and Consumer Protection, Pollution Control Team, "City of London Noise Strategy 2016 to 2026," 2016.

Coda

203 **The global market:** "Global Noise Canceling Headphones Market Is Expected to Reach $45.4 Billion by 2031: Allied Market Research," *Global Newswire*, January 26, 2023.

203 **personal sonic sanctuaries:** Mack Hagood, *Hush: Media and Sonic Self-Control* (Durham, NC: Duke University Press, 2019).

204 **"Mute promises a snake-free":** Rihard Godwin, "Hit the Mute Button: Why Everyone Is Trying to Silence the Outside World," *Guardian*, June 12, 2019.

204 **"We have been fighting":** Les Blomberg, interview by author, September 2, 2022. Unless otherwise noted, all quotations from Blomberg are drawn from this interview.

205 **A 2023 McKinsey report:** Andrea Cornell, Sarina Mahan, and Robin Riedel, "Commercial Drone Deliveries Are Demonstrating Continued Momentum in 2023," McKinsey, October 6, 2023.

205 **Wing's decision to stop:** Emmy Groves, "Drone Delivery Service Wing Stops Flying in Canberra as Business Shifts Focus to Large Shopping Centres," *ABC Radio Canberra*, updated September 18, 2023.

206 **directly beneath a drone:** Eddie Duncan et al., "Commercial Delivery Drone Routing: A Case Study of Noise Impacts," Quiet Drones international e-symposium on UAV/UAS Noise, remote from Paris—October 19–21, 2020, 132–42.

206 **keening, high-pitched whine:** Eddie Duncan, Kenneth Kaliski, and Erica Wygonik, "Three Considerations around Drone Noise," RSG (2021).

206 **In studies by NASA:** Nikolas S. Zawodny, Andrew Christian, and Randolph Cabell, "A Summary of NASA Research Exploring the Acoustics of Small Unmanned Aerial Systems," in *2018 AHS Technical Meeting on Aeromechanics Design for Transformative Vertical Flight*, no. NF1676L-27827, 2018. Andrew Christian, "Recent NASA Research into the Psychoacoustics of Urban Air-Mobility Vehicles," in *Quiet Drones 2022: Second International Symposium on Noise from UAVs, UASs, and eVTOLs*, 2022.

206 **a drone-filled future:** Los Angeles Department of Transportation, "Urban Air Mobility Policy Framework Considerations," September 13, 2021.

INDEX